SUPER
SNIPERS

SUPER SNIPERS

THE ULTIMATE GUIDE TO HISTORY'S GREATEST AND MOST LETHAL SNIPERS

**LT. COL. ROBERT K. BROWN, FORMER GREEN BERET
AND VANN SPENCER**

Skyhorse Publishing

Skyhorse Publishing books may be purchased in bulk at special discounts for sales promotion, corporate gifts, fund-raising, or educational purposes. Special editions can also be created to specifications. For details, contact the Special Sales Department, Skyhorse Publishing, 307 West 36th Street, 11th Floor, New York, NY 10018 or info@skyhorsepublishing.com.

Skyhorse® and Skyhorse Publishing® are registered trademarks of Skyhorse Publishing, Inc.®, a Delaware corporation.

Visit our website at www.skyhorsepublishing.com.

10 9 8 7 6 5 4 3 2 1

Library of Congress Cataloging-in-Publication Data is available on file.

ISBN: 978-1-5107-5544-4
EISBN: 978-1-62873-541-3

Cover design by Daniel Brount and Kai Texel
Cover photographs courtesy of Getty Images

Printed in China

To All Snipers

CONTENTS »

INTRODUCTION »

As the acknowledged number one military magazine among troops and law enforcement personnel, *Soldier of Fortune* has supported active duty soldiers and veterans for the last thirty-eight years. In so doing, it has published a multitude of articles—many of which are included in this anthology—on sniping from shortly before WWII to the present. The material is something that will be of interest to long-range shooters, snipers, and military buffs.

My "sniper credentials" are limited, since I never completed an Army sniper school. And, frankly, I don't know that I would have done so successfully, as I do not have the patience that is required of a sniper.

My first exposure to sniping came when I was assigned to be the Officer-in-Charge of the 11th Airborne Corps Advanced Marksmanship Unit (AMU) for six months prior to my reassignment to Vietnam in July of 1968. The Army actually paid me for shooting! What incredible good fortune.

In that position, I interfaced quite often with personnel from the All-Army AMU, Ft. Benning, Georgia, which was putting together the first Army sniper team that would go to Nam, commanded by Major Willie Powell. That sniper team provided the cadre for the Army sniper school in Nam. During those six months, I oversaw the training of sniper units for the North Carolina National Guard, as well as Fort Bragg.

Sniping, as a critical military skill, had been consigned to the dustbin of history by all the brass smart-asses who pompously proclaimed airpower and firepower would do away with the need for skilled riflemen after the Korean War. Major Powell had a problem in convincing company commanders that a school-trained sniper was a force multiplier; that because a graduate of the sniper school had six months combat time, they could effectively read a map (no GPS in those days), call in airpower and artillery, and was too valuable an asset to be a squad leader or charge bunkers. This problem continues even today, though to a lesser degree.

In any event, Powell's success as well as that of Major Jim Land, USMC, who was the commanding officer of the famous Marine sniper, Carlos Hathcock, guaranteed sniping would be an effective tool in future conflicts. And so it has been.

SOF's direct involvement in sniper training occurred in El Salvador, when I took numerous teams of mostly Vietnam veterans to assist/advise the Salvadoran military in their war against communist guerrillas during the 1980s. *SOF's* brilliant, long-time Small Arms Editor, Peter Kokalis, spent a number of tours training Salvo Police and "Jack Thompson," who was a Marine Nam vet, a Sergeant Major in the Rhodesian Selous Scouts and Central American contractor, conducted sniper courses for several Salvadoran Brigades, which resulted in numerous kills.

I got involved in a small way. One time, when a lull occurred in the fighting, I noted that the army units and the guerrillas stood up and yelled insults at each other about 450 yards apart. I decided I could bring some heat on the bad boys with a top-of-the-line sniper system that would reach out and touch them up to several hundred yards. Jim Leatherwood's brother, Jack, took a Winchester Model 70 action, tuned it, added a heavy Schillen barrel, and topped it off with a Leatherwood ART scope. I was hitting metal silhouette targets at Jim's Texas ranch at 1,400 meters. Unfortunately, the only time I got to draw down on guerillas, they were running through a coffee plantation. Didn't hit anyone, but I figured I made them run a tad faster.

In closing, I want to talk about Chris Kyle, one of the many participant snipers who was looking forward to being included in

Super Snipers. I only met Chris once at the 2012 Shooting, Hunting, and Outdoor Trade Show in Vegas, where he was promoting a vendor's product. I had a copy of his book, *American Sniper: The Autobiography of the Most Lethal Sniper in U.S. Military History,* and asked him to autograph it, which he kindly did. It reads:

> *SOF,*
> Thank you for all your great articles. You actually peaked my interest to join the military.
>
> —Chris Kyle

Others included in the book expressed similar sentiments. And, therefore, I take incredible pleasure in realizing *SOF* indirectly helped send scores of terrorists to Hell.

—Lt. Col. Robert K. Brown, USAR, (Ret.)

SECTION ONE: YESTERDAY'S SNIPERS

Let Our Snipers Hunt: Denying the Emplacement of IEDs

By Brian K. Sain, American Snipers Association

"The true sniper is not actually, in one sense of the term, a real 'soldier.' His nature, job, and gifts are too individualistic."

—Ion L. Idriess

These words were written nearly one hundred years ago and are still applicable today.

Unfortunately, individualism in and of itself flies in the face of contemporary military thinking; for numerous reasons, many commanding officers simply do not understand the huge force multiplier they have with their sniper teams.

U.S. Army Spc. Jason Peacock, a rifleman from Alpha Troop, 1st Squadron,
14th Cavalry Regiment, scans the rooftops from his overwatch position during
a cordon and search mission in Baghdad, Iraq, 8 Feb. 2007. The A-1/14th CAV
conducted cordon and search missions with 2nd Battalion, 1st Brigade, 6th Iraqi
National Police, to maintain security and stability in Baghdad.
(U.S. Army photo by Staff Sgt. Sean A.Foley)

The worth of the sniper in warfare has been proven time and again, but these lessons have been largely forgotten after every conflict. That is, until the newest tyrant attempts to rule the world by force and well-trained and equipped snipers are urgently needed once again. This unfortunate cycle of being "caught with our pants down" results in stop-gap measures like the designated marksman program, issuing worn out M14 rifles for precision work, and pressuring sniper schools to push students through to get enough trained snipers in the field.

Many fully trained professional snipers currently in the fight feel that the designated marksman program, with its marginal training and inadequate equipment, is "watering down" the sniper program. They feel that every soldier/marine should be a well-trained marksman first and the snipers should be left to do what snipers know how to do best. They feel it is irresponsible and downright ludicrous to send young men half trained (at best) to do a sniper's job.

The designated marksmen, however (who come from various skill levels and backgrounds), are set on a tread wheel. Commanders who do not understand sniper tactics often task the snipers. The commanders assign these men to overwatch missions, but do not properly equip them for it. The designated marksmen have no choice but to follow orders and do the very best with what they have, or with what they can scrounge, inside or outside of their chains of command. Some do a great job and eventually attend sniper school upon their return from theater.

Many snipers believe they would be better served if they had their own command and support elements rather than being treated simply as an afterthought attached to a headquarters company. But lamentations from snipers to higher echelons for better gear and operational autonomy often fall on the deaf ears of officers primarily schooled in commanding battalions of tanks, artillery, and mechanized infantry. One comment overheard from an officer defending his position was stated as "Snipers don't win wars."

That assessment may have been valid with the former "big army" threat of the Soviet Union, but may be somewhat arguable given the nature of current conflicts. Tanks and artillery are of limited use in the urban fighting of Iraq, where the number one killer of our military personnel is the improvised explosive device, better known as the IED.

IEDs are placed by human insurgents, and one of the best ways to combat them is with a corps of well-equipped and well-trained snipers. Fully supported by their command, with common sense rules of engagement, and operating upon actionable intelligence that they often develop themselves. Coalition Force snipers can methodically hunt these insurgents down and eliminate them (with no collateral damage to innocents whatsoever). The battles for Fallujah and Najaf are prime examples of virtual domination by U.S. sniper teams.

Snipers are by their very nature hunters; due to their intensive training, snipers know their own capabilities better than anyone else. The same, however, cannot often be said of their commanding officers. Many company and field-grade officers have not training in sniper employment, do not understand their snipers' capabilities, and have no idea how to deploy their snipers in a doctrinally correct manner.

Therefore, the sniper is caught in the proverbial Catch-22. Snipers require trust and autonomy from higher up in order to operate successfully (sometimes independently) and "do what they were taught to do." Unfortunately, most E5s are not able to tell a commissioned officer much about how snipers should be deployed or why the gear snipers need is different than the other troops' gear. After all, how could any E5 who has merely attended sniper school possibly know more about sniper employment than a commissioned officer who has not?

With the advent of the IED and the suicide bomber, many units that previously did not have snipers are now finding the sniper's intelligence-gathering and overwatch skills vital to mission success. Since a school-trained sniper usually carries an infantry MOS, these armor, artillery, and cavalry units are often deploying some form of designated marksman to counter current threats.

Unfortunately, the gear and weaponry required for these men may not fall within the equipment guidelines of the type of unit deploying them. For example, an armor unit may have plenty of budgetary resources for their vehicles or computer monitors for their CP, but may not have money for the optics, weaponry, and specialized equipment the designated marksmen desperately need to protect the unit from IEDs and suicide bombers. This is because their modified table of equipment may not denote that the unit even has designated marksmen!

A commonly heard reply to a sniper's request for gear and operational autonomy usually goes something like, "You guys are nothing special and no different than the rest of the troops, so quit whining."

If this is so, then why are snipers handpicked, sent to one of the most rigorous and demanding forms of training the U.S. military has to offer, and taught things other airmen, marines, sailors, and soldiers are not? Why is a designated marksman chosen for his position over someone else if everyone is equally trained and no one is special? Anyone familiar with the sniper program knows that graduates from sniper school are rarely personnel you would consider "whiners." On the contrary, snipers typically are consummate professionals and perfectionists. That is their nature and one of the primary reasons they are chosen for the job in the first place.

WORLD WAR II

Finnish Snipers in the Winter War

By John Plaster

Finland's—and probably the world's—greatest sniper of all time, Simo Hayha (left).

The One Hundred–Day War: One Shot-One Kill

During the winter of 1939–40, for some one hundred days, all alone, tiny Finland fought an immense Soviet invasion force. Out-numbered 4 to 1, the 130,000-man Finnish army and Civil Guard took on 26 Soviet divisions, tapping into their superior shooting and winter warfare skills to fight the Red Army to a standstill. Answering the challenge, *laaki ja vainaa* (one shot—one kill), Finnish snipers assisted this David-versus-Goliath fight and, in fact, one Finn scored what's regarded as the greatest number of sniper kills ever recorded.

Corporal Simo Hayha, a thirty-five-year-old Civil Guardsman from the heavily forested lake country northeast of Helsinki, shot 542 Soviet soldiers, according to several sources. A prewar competi-tive shooter and moose hunter who'd roamed Finland's woods and swamps, Hayha stood only five feet, three inches tall, but his field-craft, marksmanship, and courage more than compensated for his size.

The White Death—A Hunter's Ghost

Hayha and his comrades of the 34th Infantry Regiment performed miracles on the Kollaa Front, where Soviet soldiers trudging the deep snow often found themselves shot by "cuckoos"—snipers in trees—who deployed in fours for deadly intersecting fire and then skied away. Often operating alone, Hayha drifted like the wind, a ghost that might appear anywhere, his shapeless shadow shooting first from one flank, then skiing to fire from another direction, and then lying low to ambush the Soviets after they'd passed his hiding place. Since he was famed for his white smock sewn from bedsheets, the Soviets nicknamed him Belaya Smert (the "White Death"). Hayha averaged an astonishing five kills per day for the entire three-month war, with his highest daily score reaching twenty-five. He found no shortage of targets, especially during the Battle of Killer Hill, where thirty-two Finnish ski soldiers fought off four thousand Soviet soldiers trying to assault them in deep snow.

Hayha preferred the M28/30 bolt-action rifle, a Finnish-made, higher-quality version of the Soviet 91/30, 7.62 x 54mm Mosin-Nagant. Unlike many snipers, Hayha used iron sights, because that was how he was accustomed to shooting and because his engagement distances seldom exceeded 200 meters. Further, he believed that a scope would have raised his profile, making him more susceptible to being detected. By contrast, many other Finnish snipers had Soviet-made PE or PEM 4x rifle scopes, either purchased prewar or taken off captured Soviet weapons. Unlike the British, U.S., German, and Soviet armies, the Finns developed a special curved stripper clip that allowed these rifles to be quickly reloaded despite the scope.

Just as Dead

Some purists have questioned Hayha's score due to his choice of weapons. About half his kills were at relatively close distances, less than 100 meters, which dictated using his 9mm Finnish Soumi submachine gun rather than a bolt-action rifle. Hayha saw no such distinction, considering an invading Soviet soldier just as dead whether he shot him from ambush at 50 meters or 500 meters.

Badly wounded by a Soviet sniper on March 6, 1940, Hayha regained consciousness a week later to find a ceasefire—Stalin's bloody gambit to seize Finland had been deflected, the cost simply too great for the Red Army. In one hundred days of fighting, the Finns had lost twenty-five thousand men, but more than eight times their number—in excess of two hundred thousand Soviet Army soldiers—had paid with their lives for underestimating that tiny Scandinavian land.

The Famous Audie Murphy, Countersniper

By John Plaster

David Faces German Goliaths

Audie Murphy, America's most highly decorated serviceman in World War II, fought German snipers several times, including a personal, one-on-one fight to the death. Though only twenty years old and slightly built, the young Texan possessed a heroic spirit, backed by impressive shooting skills and an almost intuitive grasp of tactics and terrain. On October 2, 1944, shortly after Murphy had received a direct commission to lieutenant, a German sniper shot the soldier beside Murphy. The wounded man's agonizing scream attracted enemy machine gunfire, which hit a half dozen of Murphy's comrades. The very next day, another sniper fired two well-aimed shots that instantly killed two more men and halted the young Texan's unit. Murphy boldly volunteered to go after the sniper, but his company commander required him to bring along three infantrymen.

The Lone Sniper

Advancing toward the flank of the sniper's likely position, Murphy halted and ordered the three soldiers to stay there. "It was simply safer that way," he later explained. "With four men thrashing through the underbrush, the sniper would have been sure to spot one of us and perhaps kill us all." Studying where his dead comrades had been and the angles of the shots that killed them, Murphy narrowed the sniper's position to a tight sector. "Snipers always move after making a kill," Murphy thought. "Finding his new position before he spotted me was my problem." To move silently, he removed his helmet and web gear and any nonessential equipment. He carried his favorite weapon, an M1 Carbine with a fifteen-round magazine, light as a squirrel gun for quick shooting. The German sniper may have had a ballistic advantage at long or medium ranges, but Murphy intended to make this a short-range gunfight.

Audie Murphy and the German sniper rifle he captured—the hard way.
(Courtesy of Audie Murphy Research Foundation)

The Scent Of The Enemy

Inching his way along, he reached a large rock from which he suspected the earlier shots had been fired. "I sensed the presence of that sniper," he later recalled, "and he must have sensed mine." Then Murphy heard brush rattle just 20 yards away and noticed a camouflaged helmet lift ever so slowly. "He had accounted for a couple of my buddies, and I didn't feel anything as I squeezed the trigger," he recounted. "When the bullet hit him, I saw the expression on his face in the rifle sights. He didn't speak, but I had a hunch I knew what he was thinking in that last moment. He probably said in his mind, 'Lord, I am dying and I don't know why.' Then he collapsed like a rag doll and fell to the ground." Murphy seized the dead sniper's rifle and carried it back with him.

Wounded Only to Return to his Greatest Action

Three weeks later, Murphy was signaling his men by hand when, again, a sniper engaged his unit—and this time the crosshair was on him. "The bullet hit me in the right hip," he later wrote. While he lay in a ditch awaiting medical aid, the hidden sniper shot his empty helmet over and over, as if venting displeasure at only wounding Murphy. It took three months for Murphy to heal, leaving him with a slight limp. Despite that disability, he fought this greatest action on January 26, 1945, for which he would earn the Medal of Honor—after already receiving the Distinguished Service Cross, two Silver Stars, and three Purple Hearts. Murphy managed to bring his captured sniper rifle back to the States and sometimes took it out to show friends and neighbors, a fitting symbol of both his ability to outfight a dangerous foe and the nearness by which other snipers had almost taken that great hero's life.

Skorzeny's Sniper Experiment

The Legend of the Famed Commando Otto Skorzeny

By John Plaster

Ever innovative, the Third Reich's greatest commando leader, Colonel Otto Skorzeny.

A Bridgehead Across the Oder

In late January 1945, with powerful Soviet armies advancing across Poland, it looked like the Third Reich might fall within a few weeks. Already Soviet scouts had reached Germany's last natural barrier, the Oder River, only 40 miles east of Berlin.

At this critical moment, SS Chief Heinrich Himmler called on the officer he thought his country's most resourceful, the famed commando leader Otto Skorzeny, to secure a bridgehead across the Oder, at Schwedt, to threaten the Soviet armies massing near Berlin with a possible counterattack and thus buy time to improve defenses elsewhere.

Commando Extraordinary

This was quite a challenge, even for Skorzeny, Germany's "commando extraordinary." It was a division-size mission, requiring ten thousand to fifteen thousand troops, yet Himmler could provide only a handful. Skorzeny would have to find his own forces and improvise.

Improvise he did, creating infantry battalions out of thin air—everything from dockworkers in Hamburg to Luftwaffe pilots and mechanics who had no planes. From old SS friends, he borrowed an anti-tank battalion and recruited an SS unit of ethnic Germans from Romania. He assembled these and other small units around the nucleus of one SS parachute battalion—and, most unusually, one company of snipers (apparently the cadre and students from a sniper school at Friedenthal), commanded by an Oberleutnant Odo Wilscher.

Snipers in No-Man's-Land

On February 1, 1945, Skorzeny's newly named Schwedt division—swollen to fifteen thousand troops—occupied defensive positions across the Oder River. Skorzeny took special care in positioning his hundred-man sniper company, realizing "there had to be shooting at Schwedt—and accurate shooting at that." To this experienced commando leader, massing these superb marksmen at a critical position made good sense. Instead of employing snipers piecemeal, Skorzeny previously had asked generals, "Why didn't they systematically commit the snipers that each division possessed?" This he chose to do at Schwedt, concentrating all his snipers on the most critical approach route.

In his autobiography, Skorzeny explained that each night, the sniper commander, Oberleutnant Wilscher, "hid his snipers in groups of two in no-man's-land," their fire carefully integrated and overlapping for mutual support. When Soviet infantry were engaged and attempted to attack a sniper team, other hidden teams on their flanks or even their rear opened fire, confusing and repulsing the Soviet infantrymen. Wilscher also put snipers on broken-away ice sheets on the Oder River, concealing them with wood and branches. "The floating islands offered Wilscher's riflemen natural and mobile cover," Skorzeny wrote.

Snipers Buy Precious Time

Exactly as planned, Skorzeny's bridgehead drew enormous Soviet forces—he estimated his men were outnumbered 15 to 1—but they held their ground for an astonishing thirty days, a full month, and then withdrew across the Oder. Undoubtedly, this operation disrupted the Red Army's offensive timetable, buying Germany weeks to improve its defenses. As for that hundred-man company of snipers, Skorzeny concluded that these marksmen "weakened the enemy considerably. I estimated that 25 percent of our defensive success was attributable to the snipers."

KOREAN WAR

An American Sniper in Korea

By John Plaster

Above: Private First Class Henry Friday (right) lures enemy sniper fire, so Staff Sergeant Boitnott can return fire. Right: Staff Sergeant (later Technical Sergeant) John Boitnott reloads his M1C on Outpost Yoke, 1953.

Chicoms Target Outpost Yoke

To instruct the 5th Marine Regiment's newly organized sniper school, the regimental commander recruited his unit's Distinguished Riflemen and high expert shooters, among them Staff Sergeant John E. Boitnott. Considered by many to be the Corps's most accomplished Korean War sniper, this competitive rifleman—who'd earned his Distinguished Badge two years earlier—helped teach the course and then returned to the front lines to try his own hand at scoped rifle shooting.

Early on, it became a personal affair for Boitnott when a Chinese sniper's bullet ricocheted off his helmet. Clearly, enemy snipers were targeting the 5th Regiment's trenches on Outpost Yoke, but it was almost impossible to spot them. Then Boitnott devised a winning (though dangerous) countersniping technique. Partnering with Private First Class Henry Friday, Boitnott hunkered down, rifle ready, eye to his scope, while Friday voluntarily trotted along a trench line to lure Chinese fire. Sure enough, an enemy sniper rose to the bait, plinking a shot at the Marine lines—and taking in return a dead-on shot that ended his sniping career. Witnessed by Lieutenant Homer Johnson, the distance was later plotted on a map: 670 yards.

To Snare a Chicom

Over the next two days, Private Henry and Staff Sergeant Boitnott continued this tactic, resulting in nine confirmed kills at ranges up to 1,250 yards. However, when war correspondents publicized their controversial countersniping effort, higher command halted it. The 5th Regiment's 1953 staff journal recorded Sergeant Boitnott's continuing sniping:

"July 14—In mid-afternoon Sgt. Boitnott on Outpost Bruce expended one round in killing one enemy"

"July 15—S/Sgt. Boitnott on Outpost Bruce expended eight rounds of rifle ammunition in killing four"

"July 17—This morning S/Sgt. Boitnott on Outpost Bruce killed one enemy at long range with a rifle and four hours later killed another"

"July 18—S/Sgt. Boitnott of 'I' Company killed one enemy with one round of rifle fire."

As a result of his deadeye shooting, Boitnott was meritoriously promoted to technical sergeant, while reports of his countersniping appeared in newspapers across America.

VIETNAM WAR

The Marine Corps's Unrivaled Sniper

Reaches Out and Touches 103 NVA Plus 216 "Probables"

By John Plaster

A Nick Off the Right Ear Gives Him Away

Among his friends and neighbors in eastern Oregon, U.S. Forest Service officer Chuck Mawhinney blended right in. After coming home from the Marine Corps in 1970, he didn't really talk much about his service. Then, in the late 1990s, several historians and researchers discovered something quite amazing about this quiet, easygoing man. "Chuck is such a low-key and nice guy that people couldn't believe that he was capable of this," a neighbor told the *Los Angeles Times*.

To Major John Plaster

Best Regards
Your Friend
Chuck Mawhinney

Sergeant Mawhinney's Remington M40 sniper rifle, serial number 221552, is displayed today at the Marie Corps Museum in tribute to this most accomplished sniper in Marine Corps history, with 103 kills.

She was referring to the fact that her mild-mannered acquaintance was America's most accomplished Marine sniper. Not just for Vietnam, but for any war. Sergeant Charles Mawhinney had shot and killed 103 enemy personnel in 1968–69, with another 216 "probables," some at more than 1,100 yards. The only hint of such a background was a nick off his right ear, where an enemy bullet had clipped him.

Mawhinney's rural Oregon upbringing probably had a lot to do with his combat performance. Raised on a ranch, he grew up an avid hunter, trapper, and fisherman, but especially a fine rifle shot. While still a lad, he could shoot flies off a fencepost with his BB gun, so it was no surprise that he qualified as "expert" in Marine Corps Basic Training and was invited to attend Scout Sniper training at Camp Pendleton, California.

Roaming the Hills, Looking for a Fight

Arriving in Vietnam, Mawhinney was assigned to the 3rd Battalion, 5th Marines. Operating out of An Hoa, 25 miles southwest of Da Nang, his unit roamed the surrounding hills and valleys looking for a fight—and frequently found one. Rarely did Mawhinney's two-man sniper team operate independently, rather almost always in support of a company or platoon, although he might screen a flank or recon ahead of them.

"We were the company's eyes and ears," Mawhinney explained. His team often moved at night and stayed off trails and open areas to avoid booby traps and villagers who might betray their presence.

The St. Valentine's Day Massacre

Perhaps his greatest engagement took place on February 14, 1969, known among 5th Marines snipers as the St. Valentine's Day Massacre. That afternoon, while the company he was supporting dug in, Mawhinney and his teammate crept forward to watch the nearby Thu Bon River. At dusk, he mounted an AN/PVS-2 nightscope on an M14 and began surveilling the 100-meter-wide river. Soon after dark, an enemy scout waded across the river, almost stepped on the Marine sniper team, and then went back. Momentarily the scout reappeared, leading a long file of enemy soldiers, wading directly toward Mawhinney, who held his fire.

"As soon as the first one came up the bank on our side," he recalled, "I went to work. I got sixteen rounds off that night, as fast as I could fire the weapon, and every one was a head shot." Exactly like World War I rifleman Alvin York, he shot back to front so the approaching enemy could not appreciate his fire's deadliness, killing one enemy for each shot. Witnessed only by his spotter, these kills would not count toward his final tally. In fact, Mawhinney achieved 30 percent of his kills at night, exploiting night vision capability to the fullest.

Popping out of Spider Holes

"Normally I would shoot and run," Mawhinney explained. "But if I had them at a distance, I wasn't worried." That's what happened in another engagement that Mawhinney compared to shooting prairie dogs. When a Marine company came under fire, he and his partner trotted to a flank, where they could see enemy riflemen popping out of spider holes, firing, and then dropping down. He told *Los Angeles Times* writers Craig Roberts and Charles Sasser, after waiting for one to reappear, "I popped him. The lid slammed shut. To his left, another lid lifted. Another prairie dog. I popped him, too." That day he was credited with bagging three such "prairie dogs," each killed by a single shot to the head or neck.

By April 1969, Mawhinney had an astounding 101 confirmed kills of North Vietnamese and Viet Cong soldiers, in recognition of which he was meritoriously promoted from private to lance corporal. By the time he went home, he had served a full thirteen-month tour, plus two six-month voluntary extensions because, he believed, his shooting could save more American lives.

"Daniel Boone" Goes To Vietnam

Waldron, The War's Most Accomplished Unsung Sniper

By John Plaster

Staff Sergeant Adelbert "Bert" Waldron, the Vietnam War's most accomplished sniper, operated in the Mekong Delta.

Whacked from the Top of a Coconut Tree

Though relatively unknown, the most highly decorated and most accomplished American sniper of the Vietnam War was Staff Sergeant Adelbert F. Waldron of the U.S. Army's 9th Infantry Division.

In addition to being credited with 109 enemy kills—the highest count for any American sniper in any war—Staff Sergeant Waldron was also twice awarded the Distinguished Service Cross—second only to the Medal of Honor, plus the Silver Star, several Bronze Stars, and the Purple Heart.

Nicknamed "Daniel Boone" by his fellow snipers to recognize his great fieldcraft, Waldron was also a phenomenal shot, achieving incredible long-distance hits. Lieutenant General Julian J. Ewell, Waldron's 9th Infantry Division commander, recalled:

> One afternoon he was riding along the Mekong River on a Tango boat when an enemy sniper on shore pecked away at the boat. While everyone else on board strained to find the antagonist, who was firing from the shoreline over 900 meters away, Sergeant Waldron took up his sniper rifle and picked off the Viet Cong from the top of a coconut tree with one shot (this from a moving platform). Such was the capability of our best sniper.

From left to right: SFC Waldron holding an M14 with Sionics Suppressor; Sgt. Maj. Frank Moyer, author of Special Forces Foreign Weapons Small Arms Firing Technique; Mitchell Warbell IV; Gorden Ingram, inventor of the M10/M11 submachine gun.
(Photo was taken in the early 1970s at Michell Warbell III's estate in Marietta, Georgia. Photo by Robert K. Brown.)

The Darkness in the Mekong Delta Belongs to Waldron

The terrain in Waldron's area of operations especially suited sniping, with the Mekong Delta's rice paddies stretching for hundreds, even thousands, of yards, an ideal setting for long-range spotting and shooting. Equally, though, it was Waldron's exploitation of the night—employing an AN/PVS-2 night vision device and a Sionics Suppressor on his M21 Sniper System—that contributed to his great effectiveness. Many GIs in Vietnam thought the night belonged to the enemy, but in the Mekong Delta, darkness belonged to "Bert" Waldron. On the night of February 3, 1969, for instance, a staff journal noted:

> [F]ive Viet Cong moved from the woodline to the edge of the rice paddy and the first Viet Cong in the group was taken under fire . . . resulting in one Viet Cong killed. Immediately the other Viet Cong formed a huddle around the fallen body, apparently not quite sure what had taken place. Sergeant Waldron continued engaging the Viet Cong one by one until a total of five Viet Cong were killed.

Seven Rounds, Seven Kills

Time and again, he intercepted Viet Cong in the dead of night, engaging small patrols and single Viet Cong who'd mistakenly thought they could not be seen. On the night of January 25, 1969, Waldron shot seven enemy soldiers at an average distance of 350 meters—firing only seven rounds. In the darkness of January 22, 1969, he was credited with another eleven kills. On February 4, 1969, he killed another nine enemy fighters, again using a Starlight scope and suppressed XM21. Waldron was a graduate of the 9th Division's first sniper class, and General Ewell thought him the division's finest sniper.

A private man, Waldron quietly retired from the Army and, to the best of my knowledge, has never spoken publicly about his wartime service.

Silent Death

By Lt. Col. Robert K. Brown, USAR (Ret.)

Pitch black night was once again descending over the fertile, rice-producing Mekong Delta in Vietnam. As darkness moved in, it was time for Victor Charlie to move out. But tonight, the lack of

Above: Maj. Willis Powell, commander of first Army sniper team sent to Vietnam in 1968.

visibility was to prove no friend to the VC unit operating in the area, ambushed by a small element of Company D, 3/60th Infantry, 9th Division.

Six battle-hardened GIs in camouflaged fatigues positioned themselves along a tree line near a small muddy canal bordering a large rice paddy. The ambush location offered excellent fields of fire and observation of the paddy area and a road that crossed the paddy at a 45 degree angle. Once situated and concealed, the ambush party did not have long to wait.

At 1950 hours (7:50 p.m.), as dusk closed in, three heavily armed VC were observed humping down the road in an unconcerned manner about 400 meters away. In a matter of seconds, the three VC crumpled into grotesque positions of death in the red din. Shortly thereafter, a single VC emerged from a nearby banana grove. He suddenly pitched forward—dead. Two hours later, another VC peered cautiously from the woodline, attempting to determine what had happened. As he tried to retrieve the weapon and web gear from his dead comrade, he collapsed—dead.

Within an hour, another party of half-a-dozen VC moved across the rice paddy from the left of the ambush party, their deadly AK-47s at the ready. The VC point man's head exploded in a red mist and he slipped below the paddy water. The remaining VC, who were terrified because they had heard no noise, sought cover. In a few minutes, they double limed away from the dead VC. One after another, they all died, quickly and silently. At 2315 (11:15 p.m.), two more VC were observed moving parallel to the road and they died—just as quickly and just as silently. Another Charlie slopped rapidly through the rice paddy. He died, too. The "Silent Death" had struck again.

Mitch Werbell carries the M16 sniper unit with Sionics Corp. Supressor and night vision device used in Vietnam.

Eleven VC killed in three hours at an average range of 400 meters, and no noise! Did the ambush party have some fantastic new guided missile or laser beam? Negative. But they had something just as effective. A two-man sniper team armed with an accurized M14, shooting match–grade ammo, a noise suppressor (more commonly known as a "silencer"), and a Starlight scope. And the sniper credited with the eleven kills proved once again that there is nothing quite as effective as well-aimed fire for economy and results.

Since the beginning of our Army, individual marksmen have played an important part in contributing to our battlefield successes throughout years. The significance of the potential effectiveness of skilled marksmen was first impressed on European powers when General Braddock's red-coated regulars were thoroughly decimated by French troops and their Indian allies during the French and Indian Wars. Only the American frontiersmen, led by Colonel Washington, were able to prevent the defeat from becoming a total rout.

During the American Revolution, the American sharpshooter, clad in buckskin and wreaking havoc with his long-barreled Kentucky rifle, contributed greatly to the defeat of the English and their Prussian mercenaries. In fact, it was one of these gimlet-eyed woodsmen that directly affected the battle that led to the final British defeat at Saratoga.

Shortly before the Battle of Saratoga, a large unit of British infantry, led by General Frazer, who was considered one of the most effective officers in the British Army, was attacked by American forces commanded by the famed General Daniel Morgan. The British line was broken shortly after the initial engagement and General Frazer was forced to reorganize his forces.

Morgan observed General Frazer, astride his horse, directing the regrouping. Morgan sent for one of the most widely known of the famed "Morgan Riflemen," Tim Murphy, and directed him to bring down Frazer. Murphy's third shot tumbled Frazer off his horse with a mortal wound. The resulting confusion among the British unit enabled Morgan's forces to encircle the British and launch a successful attack from the rear. The loss of this position forced Burgoyne to retreat to Saratoga, where he was eventually forced to surrender.

The skills and courage of individual riflemen continued to prove of immense value to the United Stales throughout numerous conflicts. However, since World War II, the military has placed emphasis on volume of fire and fancy gadgets rather than the well-aimed shot. Generally speaking, this was especially true in the Vietnam conflict.

The Army Sniper Program in Southeast Asia was initialed in 1968, when General Ewell learned he was to assume command of the famed 9th Infantry Division in Vietnam. General Ewell believed that snipers could operate effectively in this area, inasmuch as large portions of the 9th Division's area of operations contained wide expanses of flat, cleared terrain and rice paddies. He felt there still might be a place for the individual rifleman equipped with accurate rifles and possessing patience, stamina, judgment, courage, and an eagle eye. General Ewell contacted the Army Marksmanship Training Unit (AMTU), which heretofore had been primarily concerned with developing champion competition marksmen, at Fort Benning, Georgia, and asked them if they could develop a program for training snipers fur the 9th Division.

Members of the AMTU, all of whom had had extensive rifle competition experience, enthusiastically embraced the challenge, as they were anxious to demonstrate that pinpoint marksmanship could make a contribution in modern warfare as well as punch the "X" ring out of a target. While developing a Program of Instruction (POI) for sniper training, AMTU members conducted extensive tests to determine the weapon that would best meet their requirements. The M14, highly accurized in their own shops with glass-bedded stocks, match sights, star-gauged barrels, etc., was finally selected.

The Redfield scope company was contacted to build a number of their 3x9 variable scopes with a special crosshair and stadia marks. This special Redfield scope was then married up with a special mount and a ballistic cam system developed by Captain Jim Leatherwood. This sighting system was then designated as the Automatic Ranging Telescope (ART). The mounts were designed to allow the interchanging of the ART and Starlight Scope on the same rifle, without affecting the zero of the weapon. Consequently, a sniper can utilize the same rifle in both day and night without any adjustments as long as he has both scopes with him.

Early in 1968, a silencer (more accurately described as a "noise suppressor") was added to the accurized M14 and ART, The silencer, (developed and patented by Mitchell L. WerBell III, a former OSS officer and long-time soldier of fortune, and manufactured by the Sionics Corporation, which was absorbed into the now defunct Military Armaments Corporation) reduced the noise level of the muzzle blast to the extent that beyond 100 meters, it was impossible to determine where the shot originated. Furthermore, the silencer, or noise suppressor, completely eliminated multiple flash, which proved especially valuable during night ambushes.

The suppressor increased the capability of sniper personnel, as it gave them confidence to take targets under fire that they would have normally ignored for fear of revealing their location. They found that the suppressor made by Sionics, as contrasted to other models under consideration, affected neither the range nor the accuracy of the sniper's fire.

The original sniper instruction team assigned to Vietnam consisted of one officer and eight non-commissioned officers, one of whom was a trained gunsmith. Because of the sophistication of the sniper weapons and need for minute-of-angle accuracy, the gunsmith was as important, if not more so, than any other single man on the team.

The individual sniper was to perform only the most basic maintenance on his weapon, including cleaning. About once a month, each sniper returned to the sniper school, where his weapon received a zero check and thorough cleaning. At this time, members of the original sniper instruction team determined how the equipment was performing, what problems the sniper encountered, and if the sniper was being utilized properly.

One of the main problems the sniper program faced was the misutilisation of sniper team personnel. Several snipers were wounded because they were walking point, providing rear security, or assaulting a fortified position. Many small unit commanders attempted to

An Australian officer tries his hand with the supressed M16 at a demonstration near Saigon in 1969.

get sniper personnel transferred into slots where they would serve as squad leaders or platoon sergeants. Company and Battalion Commanders, who were not familiar with the sniper's capabilities, were aware, however, that said personnel were highly trained, motivated, and competent in a wide variety of military subjects and, furthermore, had extensive combat experience.

When the first sniper instruction team arrived in Vietnam in June 1968, they found that they would have to build their own range and instruction facilities. At that time, the only accurized M14s with AKs were the ones brought by the instruction team from Fort Benning.

While the range was under construction, the instructors themselves operated as snipers with units in the field in order to determine what problems their future students would have to cope with and to familiarize small unit commanders with their unique capabilities. Concurrently, they also organized a familiarization and zeroing program on the M16 for replacements coming into the 9th Division.

Sixty-five of the accurized M14s and sophisticated scopes arrived in Vietnam in late October 1968, and the snipers began dealing large doses of "silent death" to the unsuspecting Viet Cong and North Vietnamese regulars.

In selecting students to participate in the sniper school, it was determined that a man should have at least three months combat experience with a field unit, that he should have at least six months remaining in Vietnam and, above all, that he must volunteer for the program. Initially, unit commanders sent men to the school who were not volunteers—who had no idea why they were there or for what. These individuals were quickly identified and all returned to their units.

The sniper school lasted eighteen days and started with a review of basic marksmanship fundamentals, such as positions, use of slings, reading and hold off for wind, care and maintenance of the weapon, plus other related subjects, such as target detection, map reading, land navigation, camouflage, tactics, adjusting artillery, etc.

On the average, only 48 percent of a thirty-man class was graduated. Some dropped out because they were not willing to put forth the necessary effort; others, when it became apparent they could not shoot accurately and consistently. It was possible to score 160 points in the two field firing courses. The wash out score was 120, and no exceptions were made for those who failed to make it.

Tactical employment of snipers varied. Trial and error, coupled with flexibility, allowed the sniper units to develop their own particular brand of tactics that produced the most kills. Initially, sniper teams were employed with Intelligence Squads and with Battalion blocking forces with little success. Then sniper teams were inserted with company-size units on the evening resupply chopper. This tactic produced good results until the VC wised up and began giving company size units a wide berth. Eventually the sniper teams found the most effective tactic was to work on their own with a four-man security element. This was the tactic used to kill the eleven VC noted earlier.

In any case, results proved that the tactics selected worked and that there still is a place in modern warfare for an accurate shot. From January 7 to July 24, 1 percent was shot, for instance. U.S. Army snipers in Vietnam accounted for 1,245 VC, expending an average of 1.37 rounds per kill!

One of the unsung heroes of the Vietnam conflict is the sniper who accounted for the eleven VC mentioned previously. He is Sfc. Adelbert F. Waldron (Ret.), who during his tour as an Army sniper was credited with 113 confirmed VC kills and ten blood trails. And in only eight months!

Waldron, who was one of the first snipers graduated from the new sniper school, found that, initially, he spent as much time selling his program to skeptical commanders as he did killing VC.

For his first six missions, he lay on the floor of a Huey chopper, scanning the countryside through his Starlite scope for wandering VC. When he picked up a VC who thought he was safe in the dark, Waldron would fire tracers to direct the gunship to the target. Finally, he started operating with small ambush parties and, as his body count grew, he was dubbed with the code name, "Daniel Boone." On one occasion, he dispatched a nine-man VC column, working from rear to the front. With his weapon silenced, the VC were unaware that they were being eliminated one by one.

Below are a few after-action reports describing the activities of Sfc. Adelbert F. Waldron (Ret.):

Sergeant Waldron and his partner occupied a night ambush position with Company A, 3/60th Infantry on January 25, 1969, approximately four kilometers southeast of Mo Cay (XS 502178). The area selected for the ambush was along a dike adjacent to a rice paddy. The configuration of the terrain was such that a firing platform had to be improvised from C-ration boxes in order to provide clearance over the paddy dike. After the sniper position was completed, nipa leaves were used to conceal the team. At 1910 (7:10 p.m.) hours, two Viet Cong moved from right to left across the edge of the paddy and Sergeant Waldron engaged both, resulting in two Viet Cong killed. A half hour later, three more Viet Cong crossed. The same area and Sergeant Waldron took them under fire, resulting in three more Viet Cong killed. The next contact took place at 2232 (10:32 p.m.) hours, when one Viet Cong returned across the paddy, apparently in an attempt to extract the weapons or bodies of the fallen. Sergeant Waldron subsequently engaged and killed this Viet Cong. The final contact of the night came at 2355 (11:55 p.m.) hours. A single Viet Cong came out into the paddy, was taken under fire, and killed. A total of seven enemy soldiers were killed at an average range of 350 meters with a total of seven rounds having been fired.

Sergeant Waldron and his partner occupied a night ambush position with Company D, 3/60th Infantry on January 22, 1969, approximately 7 kilometers south of Mo Cay (XS 480128). The night ambush sight was along a tree line near a small canal, bordering a large rice paddy. The snipers position offered them very good fields of fire, including excellent observation of a road which crossed the paddy at a 45 degree angle. At 1950 (7:50 p.m.) hours, three Viet Cong were observed walking down the road, and they were taken under fire by Sergeant Waldron and his partner, resulting in three Viet Cong killed. Only one of these kills was credited to Sergeant Waldron. Shortly thereafter, a lone Viet Cong came out of a banana grove and was taken under fire by Sergeant Waldron, resulting in one Viet Cong killed. The next contact took place at 2100 (9:00 p.m.) hours, when one Viet

Cong appeared from the woodline, attempting to retrieve the weapon and web gear from the fallen body. The Viet Cong came out of a banana grove and was taken under fire by Sergeant Waldron, resulting in one Viet Cong killed. Within an hour, a group of five or six Viet Cong began moving across the rice paddy from the sniper's left. Sergeant Waldron took the group under fire, killing one Viet Cong and causing the others to drop to the ground. After a few minutes, the entire group got up and continued moving and were subsequently engaged one at a time until a total of five Viet Cong were killed. The next contact took place at 2315 (9:15 p.m.) hours, when two Viet Cong were observed moving near the road. They were immediately engaged by Sergeant Waldron, resulting in two Viet Cong killed. Approximately ten minutes later, Sergeant Waldron's position began receiving probing fire from an AK-47. Continuing to scan the area with his Starlight scope, Sergeant Waldron spotted one Viet Cong running across the rice paddy and he was immediately engaged by Sergeant Waldron, resulting in one Viet Cong killed. The probing fire continued and therefore the sniper team moved to an alternate position for the remainder of the night. A total of eleven kills were credited to Sergeant Waldron during the night.

Sergeant Waldron and his partner occupied a night ambush with Company B, 3rd Battalion, 60th Infantry on 30 January 1969, northeast of Ben Tre (XS 528351). The area selected for the ambush was an intersection of two dikes surrounding a large rice paddy. The fact that the rice had recently been cut provided the snipers with good fields of fire and enabled them to use a prone position. Just before dark, two or three individuals were sighted moving towards a nearby village. Curfew was not in effect at that time and therefore the individuals were not fired upon. At approximately 2000

(8:00 p.m.) hours, one Viet Cong was observed moving near a tree line forward of the snipers' position and a request for artillery fire was called in. The request was denied since the area was considered populated. Sergeant Waldron observed the Viet Cong again and engaged him, resulting in one Viet Cong killed. The next contact look place at 2040 (8:40 p.m.) hours, when sixteen Viet Cong were observed moving in a line across the edge of the rice paddy. Sergeant Waldron took the first VC under fire resulting in one Viet Cong killed. The remainder of the group immediately hit the ground. Five minutes later, the group got up and resumed moving, apparently not sure of what had happened. Sergeant Waldron engaged and killed one more Viet Cong, causing the remaining Viet Cong to panic and start running towards the ambush position. They apparently thought the fire was coming from the woodline. Sergeant Waldron subsequently engaged and killed five more Viet Cong, bringing to eight, the total number of Viet Cong killed during the night. Eight rounds were fired in obtaining these kills at an average of 500 meters.

Sergeant Waldron and his partner occupied a night ambush with Company D, 3rd Battalion, 6th Infantry on 3 February, 1969, approximately three kilometers south of Ben Tre (XS 518281). The area selected for the ambush was in a large rice paddy bordered by a wooded area. At 2109 (9:09 p.m.) hours, five Viet Cong moved from the woodline to the edge of the rice paddy and the first Viet Cong in the group was taken by Sergeant Waldron. The first shot missed the target, necessitating that Sergeant Waldron readjust his Starlight scope. The missed target prompted his partner to comment, "You missed that one, didn't you." After necessary adjustments were made, Sergeant Waldron again engaged the first Viet Cong in the group, resulting in one Viet Cong killed. Immediately the other Viet Cong formed a huddle around the fallen body, apparently not sure of what had taken place. Sergeant Waldron continued engaging the Viet Cong one by one until a total of five Viet Cong were killed. The next contact look place at 2225 (10:25 p.m.) hours, when one Viet Cong returned across the rice paddy, apparently looking for equipment and weapons near the bodies of the fallen Viet Cong. Sergeant Waldron took him under fire, resulting in one Viet Cong killed, bringing to six the number of Viet Cong killed during the night.

Sergeant Waldron and his partner occupied a night ambush position with Company D, 3/60th Infantry on February 4, 1969, approximately three kilometers south of Ben Tre (XS 527283). The area selected for the ambush was at the end of a large rice paddy adjacent to a wooded area. Company D, 3/60th Infantry, had conducted a MEDCAP and 1CAP in a nearby hamlet during the day, hoping to gain information on Viet Cong movements in the area. At approximately 2105 (9:05 p.m.) hours, five Viet Cong moved from the wooded area toward Sergeant Waldron's position and he took the first one in the group under fire, resulting in one Viet Cong killed. The remaining Viet Cong immediately dropped to the ground and did not move for several minutes. A short time later four Viet Cong stood up and began moving again, apparently not aware of the fact they were being fired upon from the rice paddy. Sergeant Waldron took the four Viet Cong under fire, resulting in four Viet Cong killed. The next contact took place at 2345 (11:45 p.m.) hours, when four Viet Cong moved into the rice paddy from the left of Sergeant Waldron's ambush position. The Viet Cong were taken under fire by Sergeant Waldron, resulting in four Viet Cong killed. A total of nine enemy soldiers were killed during the night at an average range of 400 meters. Sergeant Waldron used a Starlight scope and noise suppressor on his match-grade M14 rifle in obtaining these kills.

As his body count rose, so did his fame—among both American units and the enemy. Victor Charlie quickly tired of Waldron's game rules and put a $50,000 price on his head—dead or alive. Once American intelligence found out about the price on Waldron's head, he was hustled out of the land of rice paddies and "silent death" in twelve hours!

Sixty-five accurized M14s, with suppressors and adjustable ranging scopes (developed by SOFer Jim Leatherwood) arrived in Nam in October 1968.

Apparently General Ewell was pleased with the sniper program, as he noted in *Impressions of a Division Commander in Vietnam,*

The most effective single program we had was the sniper program. This look a whole year to get off the ground from scratch, but we ended up with eighty snipers who would kill (or capture) from 200 to 300 enemy per month. Not only did we get this direct return, but they also encouraged the other men to shoot well. Snipers, like everything else, are highly-sensitive to tactics and techniques, so one has to handle them well. The flat, open delta terrain was ideal for snipers. Other divisions are now trying snipers in other areas, so we shall see how they work on a broader basis. Snipers had been tried before in the theatre with tepid results, but we insisted that the program be exactly right, demanded results and got them.

Daring his tour of duty, Waldron became one of the most highly decorated soldiers of the Vietnam conflict. He was awarded three Distinguished Service Crosses, two Silver Stars, four Bronze Stars, three Air Medals, and two Purple Hearts.

In light of all the controversy surrounding the accuracy of "body counts" reported by both the South Vietnamese and American units, it is worth noting that in the course of calculating sniper kills, no VC or North Vietnamese was counted as a "kill" unless an American trooper, either the sniper or a member of the support unit, actually was able to physically place his foot on the body.

Wars may come and wars may go, but the valiant U.S. Army sniper personnel again proved the value of a man and a rifle. Let's hope the Army doesn't forget this lesson as it has others in the past.

SECTION TWO: SNIPERS DURING THE WAR ON TERROR

5

Ultimate Snipers

By Mark Gongea

"Sent it!" Acting as spotter for USMC sniper vet Steven Reichert, veteran Green Beret John Plaster focuses on a distant target.

In recent conflicts in Iraq and Afghanistan, the majority of U.S. and Coalition casualties have come from improvised explosives devices (IEDs). The insurgents figured out early on that they could not fight the U.S. military in gun battles, so they relied on proven guerrilla tactics.

But the terrorists' asymmetric advantage lasted only until the Coalition forces came up with countertactics. One important game changer was the rapid improvement in weapons, equipment and employment of snipers. As the United States and its allies improved the technology of sniping and, perhaps more importantly, the use of snipers, success stories of astounding feats—shots at ranges previously thought impossible, under difficult conditions and sometimes through unlikely objects, such as brick walls—started coming back from the front.

In 1993, Paladin Press advanced the training of snipers with the publication of *The Ultimate Sniper* by Major John Plaster (revised and updated in 2006). That book was followed by two *Ultimate Sniper* videos, the last one released in 1996. In view of developments on the battlefields of Afghanistan and Iraq, it was clearly time to take another look at snipers, their weapons, their technology, and their employment.

There could be no better person to take on this task than John Plaster, the author of the *Ultimate Sniper* books and videos and Special Forces combat veteran. To help him with this modern update on today's snipers, Plaster enlisted the help of Canadian Robert Furlong, who holds the world record at more than 2,700 yards for a confirmed kill with a .50 caliber rifle; U.S. Army First Sergeant James Gilliland, who holds the record for the longest conventional sniper rifle kill in Iraq; and former U.S. Marine sniper Steve Reichert, who achieved the longest .50 caliber kill in Iraq and famously fired through a brick wall to eliminate a machine gun team endangering his Marine company.

Whenever you have the opportunity to be around the top people in any field for even a brief moment, you are indeed fortunate.

I had the chance to spend several days watching and listening to the premier instructor of snipers and three of the top snipers in the world demonstrate the state of today's art of sniping. *The Ultimate Sniper III* video shoot promised to be a revealing look into this world.

* * *

Spring usually arrives late in northern Wisconsin, but this year it had been exceptionally tardy. When the Paladin video crew arrived in early June to film the third installment in the Ultimate Sniper video series in Iron River, Wisconsin—the home of Major John Plaster—the snows from early May had barely melted away. The lingering cool temperatures and moisture had left the area green and lush and perfect for shooting a sniping video.

The Paladin crew consisted of publisher Peder Lund, video director Matt Doyle, cameraman Brad Efting, and me, a freelance video editor from Canada, who was participating in my first Paladin video shoot. We arrived on Thursday afternoon and were scheduled to meet Steve Reichert at the airport in Duluth, Minnesota, for the 54-mile drive to Iron River. At the airport, we learned that Reichert's flight into Duluth had been delayed and he would not be arriving until very late that night. So we drove to Iron River, confident that the Marine sniper could find his way to town. Neither Doyle nor I had met John Plaster before, and we were curious about Paladin's best selling author. The first thing we learned about the retired U.S. Army major who spent three tours in Vietnam with MACV–SOG was that he doesn't waste time with casual talk.

Reconnaissance

"Let's go and check all the locations," he told us as soon as the introductions were completed. "Here are a list, a map, satellite snapshots, and characteristics of each place," he announced, passing out copies of his files to each of us. We were all impressed by Plaster's preparations. He had delivered the detailed outline weeks ago, and the

Former Canadian sniper Robert Furlong takes aim with a Barrett .50 caliber sniper rifle. In 2002, he set a world record for a .50-cal kill, dropping a Taliban leader at 2,700 yards in Afghanistan.

handout included maps, list of locations with all data we'd ever need, satellite pictures with measurements, a summary script for each scene, plus detailed descriptions of locations and instructors.

"Great," I thought to myself, "we just have to set up the gear and push the record button." That turned out to be not quite true.

"We're going to use my truck," Plaster instructed. "It's better to ride together so we can talk about the shoot."

All five of us crammed into the SUV and were soon racing to the first location, the "241" point, as Plaster had named it on his hand-drawn map. During the 10-mile drive, deep in the Chequamegon National Forest, we had time to admire the wildness of this place. There were only woods as far as the eye could see.

"We are in Bayfield County—twenty thousand people and not a single stoplight in the whole county," Plaster told us as he pulled over on the side road. "This is 241, where we're going to fire in extreme-range conditions, over 1,000 yards."

Everything around was charred, the aftermath of a massive fire. "It was a controlled burn," Plaster explained as we looked around at the terrain, made a few notes about staging shots, and asked a few questions before heading out for the next location. "Wyoming," he called it.

"The most important objective for this shoot is to find safe locations, where we can send rounds at long distance. I personally selected all of them. I got lucky with my friends; a lot of people were offering their help to make this video a success," offered Plaster.

In the next hour, we saw the last two locations, "1800" and "The Sheriff's," plus two backup locations in case of bad weather. All were perfect.

"Ready, Aim, Shoot!"

After a very successful first day, we were eagerly looking forward to Friday, when we would meet the three decorated combat snipers. Day Two started with a strong handshake delivered by a very strong

A U.S. Army sniper team, armed with an M107 Barrett .50 cal. semi-auto rifle, prepares to engage a distant Taliban fighter in Afghanistan.

Marine. Steve Reichert had arrived late in the night, driving from Minneapolis, but he appeared rested and ready to start filming.

We loaded up and headed to "1800" for the first scenes to be shot with Reichert. As Steve was getting his sniper rifle ready, John Plaster took snap shots with a Savage TRR-SR .17 cal Hornady Magnum Rimfire rifle, and explained the unexpected choice of rifle and ammo.

"This ammo is not expensive, but it is high quality, perfect for snap-shooting and positional shooting, where a lot of expensive match-grade ammo could be consumed. Use a rim fire to get the desired results before you start training with a combat rifle."

Though he is only thirty-one, Reichert has a long history as a sniper. It started when he was in the sixth grade and bought John Plaster's *Ultimate Sniper* book from Paladin Press and read it cover to cover, over and over. Hiding in a tree house, he would pretend that he was a Marine sniper in combat and use the book's information to decide what to do in various scenarios.

"Soon after that," Reichert said, "I bought the *Ultimate Sniper* video with all the money I could save from my chores, especially shoveling snow. It was at that time I decided what path I would follow: join the Marines and become a sniper."

Sniper Jim Gilliland steps upon his spotter's back to gain enough clearance to take a shot in Ramadi, Iraq. Gilliland's team was credited with more than two hundred enemy kills.

Later that afternoon, Jim Gilliland and Rob Furlong joined us, and the discussion among the snipers soon turned to the differences between professional soldiers and insurgents.

"The enemy soldiers have no morality, no ethic. They have never heard of Geneva Conventions, and even if they had, they will never comply with its content," remarked Reichert of his observations from Iraq.

Gilliland, who was an instructor at the U.S. Army Sniper School and whose "Shadow Sniper Team" was credited in a USA Today article with more than two hundred kills, shared two of his combat stories to illustrate how the enemy isn't bound by the same rules as Western forces.

"They can operate in mosques, churches, and hospitals, and we can't really go in there. They can even dress up as women. Late one afternoon in Ramadi, one of the shooters and I were observing an area, and we noticed a woman dressed in a complete burqa come out of a side street and very near to a known IED location. Her movements were very erratic, a lot of looking around, and she just did not fit in. Basically, she was out a lot later than we were used to women being out. So we watched this individual for about twenty minutes until she turned and walked back into the alleyway. The two of us talked back and forth, and Harry and I decided this was something we really needed to pay attention to. Then the individual came back out, looked around a time or two, and then immediately bent over where the IED hole was and started brushing material out of the way.

"That gave us what we needed to positively identify hostile intent. So I got on the spotter scope, and Harry jumped on the rifle and with a 700-yard one-round kill laid this individual over. We were very nervous about this kill because we didn't know what the reaction was going to be since the target was a woman. The two of us talked it over and spent the night doing some reports to preempt what was might happen.

"The next morning when the sun started coming up, we both noticed that the woman was still lying there, which was absolutely not normal. Usually the bodies were policed up during the night and as a general rule buried before the sun came up the next day. But this individual was still out there. I got back on the spotting scope and looked down at the body. The burqa face cover had flipped up, and underneath was a large Wahabbi (fundamentalist) beard. It was actually a man who had come in and started digging a hole with the intention of planting an IED.

"Another time," Gilliland continued, "we had an individual, a male, who had been shooting at one of our patrols, put a child, four or five years old, on his shoulders to give him a piggyback ride. But the man was still walking around with his weapon, thinking that no one would engage him because of the child. Fortunately for the troops who were on the ground, we had a sniper team right there. It was a 200-meter shot, very low risk, so we put three rapid rounds into the individual to take him out. The child ran off unharmed.

"These two scenarios we just talked about give you an idea of what you might see on today's battlefields. You are not going to fight against a uniformed individual; you are not going to fight against a soldier or a warrior. You are going to fight an indigenous person using the things they have and know, and they are willing to do any-thing to win."

Gilliland holds the record for the longest sniper kill recorded for a 7.62mm rifle in Iraq, shooting into the fourth floor of a hospital in Ramadi at 1,250 meters to eliminate an enemy sniper who'd just killed an American. He took this difficult shot with the scope's turret set for maximum elevation and with the scope's mil-dot reticle held over an additional 2 mils, or approximately 12 feet of "Kentucky windage."

"It was a one-in-a-million shot that I do not expect to be able to duplicate," added Gilliland modestly.

Extreme Range Firing

On our second day of shooting we moved to "241," where we would be filming heavy rifle live fire at extreme range. We noted that it was cooler than the previous day, but temperature was of little concern to the snipers.

U.S. Navy SEAL Lieutenant Michael P. Murphy, right, with fellow SEALs in Afghanistan shortly before his June 28, 2005, mission for which he was posthumously awarded the Medal of Honor. The Ultimate Sniper III *video is dedicated to him and another qualified sniper, Staff Sergeant Robert J. Miller, a U.S. Army Special Forces posthumous Medal of Honor recipient.*

"The wind is a sniper's biggest challenge," Plaster explained. "A wind gauge only tells you the wind where you are; it cannot tell how the wind is behaving where your target is, and it cannot give you an estimate for the wind along the projected trajectory of the bullet. You cannot predict the wind's direction and velocity unless you see its effect on trees or other natural elements."

Reichert elaborated. "Snipers these days carry a lot of technology with them. In the past, you might have had a thermometer, a compass, your data sheet and log book, and that would allow you to make an educated decision on a firing solution with what you had in front of you. Today, shooters have a weather station that will give barometric pressure, temperature, and wind speed. That's great, but it's what you do with the information that counts, and that's where PDAs [personal digital assistants] come into play.

"I can plug all the environmental data into the PDA to produce a firing solution that will get me a lot closer to the target on that first

The new sniping DVD is dedicated to two sniper-qualified posthumous Medal of Honor recipients, U.S. Army Special Forces Staff Sergeant Robert J. Miller (shown) and U.S. Navy SEAL Lieutenant Michael J. Murphy.

round. When you are shooting at extreme distance, figuring out the trajectory isn't necessarily the hard part—figuring out the wind is. With a PDA you can enter multiple winds in different zones. If I have a wind going one way here and a wind 500 meters away going a different way—wind going in different directions all the way to the target—I can plug those into the PDA to get a firing solution to get the first round on target with a lot greater efficiency and accuracy."

Reichert assembled his .50-caliber rifle, an impressive M107 Barrett, which was being used for the first rounds at more than 1,000 yards. The ammunition was Hornady AMAX .50 BMG match with a 1.050 ballistic coefficient and premium propellants. Rob Furlong would be assisting Reichert.

Furlong explained his role. "When shooting at extreme range, it is very important to have a good spotter. The spotter will look through

The new Paladin Press instructional DVD features four world-class snipers: (L—R), Canadian world record .50 cal. shooter, Robert Furlong; U.S. Army Special Forces SOG combat veteran, John L. Plaster; the USMC sniper with the longest record kill in Iraq, Steven Reichert; and U.S. Army Airborne Ranger sniper, Jim Gilliland, who achieved the longest 7.62mm kill in Iraq.

his scope and will watch the arc of bullet trajectory as it transits mirage or rippling heat waves. The bullet makes a 'swirl,' and that is what a spotter carefully observes. He can also see how and where the wind will change the trajectory, the wind intensity and direction, and therefore he will be able to make corrections and assist the sniper in making the adjustments. The best spotter is a former sniper, with years of experience in watching the bullet in the air."

"How can you see that swirl, Rob? How does it look?" someone asked.

Rob smiled. "It is like in the Matrix movie. Remember when everything is in slow motion and you can see trajectory of the bullet in the air? Also, another important element to consider is the 'splash' the bullet makes when it hits either the ground or another surface."

The cameraman set one of the cameras, which was positioned close to the target, to record and then backed away to a safe distance. We were around some of the most precise shooters in the world, but we were dealing with deadly rounds so all safety precautions were observed.

Furlong looked over the valley to the target. "In extreme-range shooting, it is too far to estimate by eyesight. I'm curious to see how it is going be here in the valley. The wind is going be challenging, as always."

From his prone position, Reichert interjected: "At extreme range, your aim is to put enough rounds out there that you're bound to hit something."

The .50 caliber slammed into Reichert's shoulder, and the bullet travelled the 1,000 yards, its swirl and splash relaying precious information to his spotter, Furlong.

"Two mil on vertical, 1 mil right, Steve."

"Okay, Rob, ready to fire!"

"Send it!"

"Windage is good, 1 mil down."

"Up!"

"Send it!"

Another bullet swirled through the air. This time, Rob watched the impact with the steel target.

"Hit!"

Lying nearby behind a Savage .338 Lapua Magnum equipped with a Millett LRS 6–25x56 scope, John Plaster placed a first-round hit into a target at 744 yards. Adjusting his scope, Plaster repeated that at 1,000 yards with another first-round hit. Then, firing the same rifle, Furlong cranked up the elevation and hit another steel target, well beyond 1,000 yards. That was less than half the distance of his longest confirmed sniper kill in Afghanistan (2,700 yards), which remains the world record .50 caliber shot. In 2009, a British soldier, Corporal Craig Harrison, exceeded Furlong's shot by a mere 100 feet—but that was with a .338 Lapua Magnum, not a .50 caliber.

While waiting to fire, Furlong told us about his confirmed kill in the rugged mountain terrain of Afghanistan in March 2002, while he was attached to a U.S. Special Forces team.

A pair of rooftop U.S. Army snipers covers a passing American convoy in Baghdad. The nearer sniper's weapon is an M24 bolt-action rifle, while his partner (background) has an accurized M14 designated marksman weapon.

"A group of three al-Qaeda fighters were moving into a mountain-side position in Shah-i-Kot valley at about 9,000 feet above sea level. They were walking, maybe thinking it was their lucky day, but . . . it wasn't," he emphasized.

Furlong described the conditions that surrounded his shot. Previously, he had run out of Canadian ammo and was firing U.S. ammo, which was "hotter," which means it traveled flatter and farther. Approximately 1½ miles (2,700 yards) away, he knew that his long-range sniper weapon (LRSW), a .50-caliber McMillan TAC-50 long-range sniper rifle and ammunition loaded with 750-grain Hornady A-MAX very-low-drag bullets, had a maximum effective range of 2,190 yards. Taking advantage of an old sniper trick, Furlong laid out his ammo in the sun to warm it, hoping to coax even more distance from the rounds.

Furlong also had to adjust for elevation drop, as well as the three crosswinds swirling at the time. "I was maxed out on both elevation drop and windage," he remembered, "so I halved the scope and led the target four mils for windage and four mils for elevation. I took aim at a Taliban fighter carrying an RPK machine gun. The first round missed the target; the second hit the knapsack on his back; the third struck the target's torso, killing him."

With a muzzle speed of 823 m/s (2,700 feet per second), each shot reached the target approximately four seconds after being fired. "A .50 round is devastating. If a .50-caliber round hits you, you're not going to live to tell about it. I knew I hit him, and that was part of my job, to eliminate the enemy threat," Rob concluded and assumed his position to fire the first round for the camera.

Ultimate Snipers, Part II

By Mark Gongea

"Death From Above"

Back at "1800" again, the snipers prepared to demonstrate up/down slant firing from a 34-foot tower, then concrete-wall penetration, and finally disabling an automobile with sniper fire.

Up in the tower, Reichert laid prone with a .338 Lapua Magnum. Furlong assisted him on the scope, and Plaster observed the target through binoculars. The target was a 10-inch steel plate at 450 yards. After the first three rounds, Reichert was surprised that Plaster hadn't confirmed a hit on target. Furlong started doing the math again.

"How did you zero the rifle, Steve?" Plaster asked.

"I did it without the suppressor. Do you think that is the problem?"

"Maybe it's better to take it off."

Reichert removed the hot suppressor. After another five rounds, Reichert shook his head, apparently disagreeing for the first time with his comrades. The bullet's splash was confusing them. Some hits looked like the aim was too low, some like the aim was too high, and after all the adjustments were made, sometimes the splash was right behind the steel target.

"I think we're chasing it," Furlong offered. "We cannot hear the steel sound, but it looks like you hit it at least three times."

Later on, the tape from the close camera confirmed that Reichert had hit the target six times—his powerful .338 Lapua Magnum bullets passing completely through the steel target!

Hitting a Brick Wall

Still at "1800," the snipers set up for another live-fire demonstration. Furlong and Reichert were prone, both with .338 Lapua Magnums, while Gilliland took his position in the tower with a .50-caliber

M107 Barrett. Their mission was to neutralize a potential threat concealed behind a concrete block wall.

At Plaster's command, all three snipers fired. From a safe distance, we witnessed the wall's destruction. In less than two minutes, the three-foot-high wall was smashed to pieces. Two silhouette targets behind the wall were struck by bullets in lethal parts of the body. But more importantly, block fragments, too, had damaged and torn the targets, demonstrating that secondary debris would have killed or wounded a hidden gunman.

Reichert could attest to this because he had actually done this in Iraq in April 2004. Firing his M82A3 .50-caliber at the enemy, at a distance of 1,614 meters, Staff Sergeant Steve Reichert saved the lives of his fellow marines and also took out three enemy gunmen when his bullets smashed through a brick wall.

As fourteen Fox Company Marines patrolled the road into Luta-fiyah looking for roadside bombs, Reichert and his spotter were atop an oil tank about 1,000 yards away, providing cover. The sniping team spotted what they thought could be IED wires from a dead dog lying in the road. Reichert radioed the marines down below and asked them to check the carcass. The marines found the IED that had been emplaced in the dog and were about to dispose of it when they began taking RPG and AK fire from insurgents in the town. From their vantage point above, Reichert's team could see the fire-fight unfolding, and they could clearly see that their marines were pinned down and taking fire from a machine gunner on the rooftop of a nearby building. Taking aim at about three-quarters of a mile, Reichert fired his first shot low. He then "doped his scope," based on feedback from his spotter, and fired a second shot, taking out the enemy gunner.

But that wasn't the end of the battle. Reichert and his spotter saw three insurgents with a machine gun climbing some stairs to take refuge behind a brick wall. Knowing that they must take them out before the insurgents got the gun operational, he and his spotter calculated for the shot that had to penetrate the brick wall and incapacitate three combatants from a mile away. Reichert steadied for the shot, pulled the trigger, and watched his armor-piercing Raufoss Mk 211 round penetrate the wall. The shot killed one insurgent, and fragmentation from the bullet and the bricks likely injured the others.

Reichert manned the .50-caliber rifle for twenty hours before the attack was over. For his actions that day, he received the Bronze Star for Valor. In the after-action report, the platoon leader stated that Reichert had made his record kill shot from 1,614 meters—just over a mile away. Reichert's accuracy was cited as the deciding factor in the outcome of the firefight.

"I was concerned about my marines making it out of there in one piece," explained Reichert about his motivation to take the difficult shot.

"The lesson I learned that day in Iraq was that, no matter what, you keep going. Never quit, and you can pass any obstacle."

Full Auto

The snipers next turned their deadly fire to an automobile engine. From our position about 400 yards away, we couldn't see what the impact was, but when we got closer, we noticed that the engine had been hit repeatedly in the most critical parts. One round penetrated 2½ feet inside the engine, while another severed the oil dipstick, shattering the crankcase. No mechanic in the world was ever going to make this engine run again. The rounds were so precise

Rob Furlong with a McMillan TAC-50 similar to the TAC-50 he used to execute the 2,430 meter (2,657 yards) shot in Afghanistan during Operation Anaconda in March 2002.

that everything else on the vehicle was still intact; the damage was confined to the engine area.

Sniping Round Table

One of the most interesting parts of the shoot for me occurred on Sunday and involved no guns, shooting, or cameras. Just four snipers sitting around a table chatting about sniping, combat, and their personal experiences.

As the snipers convened at the table, Furlong approached Plaster holding a copy of *The Ultimate Sniper* and asked him to autograph it. Furlong held a copy of the original 1993 edition, and it was beat up and well worn, like it had been involved in a few skirmishes itself.

"Let me give you a new copy of the updated *Ultimate Sniper*," offered Plaster.

"No," Furlong replied, "this book traveled with me to Afghanistan. I would like you to sign it."

Gilliland added that most snipers he had met had taken their

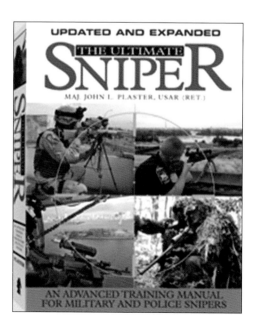

copies of *Ultimate Sniper* to war with them in Iraq, Afghanistan, or wherever. Plaster autographed the battle-proven book, but also gave Furlong a signed copy of the current edition.

What Role Does Luck Play?

To start the discussion, Plaster said, "All of you are record shooters. Jim, yours was the longest .308 shot in Iraq. Rob, yours was the longest shot in Afghanistan. And Steve, you received the Bronze Star for your sniping in Iraq. And Rob bested Carlos Hathcock's record that stood for thirty-one years. To what extent does luck play a part in such shots?"

"Luck does play a part," Gilliland conceded, "but knowledge is the single-most important thing: knowing your equipment and your own capabilities."

"In other words," Plaster remarked, "it's all luck—the more I practice, the luckier I get. Your knowledge and skill make such shots possible but not probable."

Of course, there is always bad luck in war, as well. No matter what you do, or don't do, as a sniper, you still take losses.

"It's probably 95 percent skill, training, and preparation, but there is also the five percent that's bad luck," observed Plaster. "A guy stands up at the wrong time and gets hit."

"It's a lottery you don't have to be present to win," Gilliland confirmed.

"For your record shot, Jim, it was a friend you were defending, correct? Did you realize that at the time?" asked Plaster.

"We did. We knew who was up there taking fire. We had built a good relationship with everyone in the company, not just the leadership but with the sergeants and the specialists. We got them to recon areas that we were thinking of using for sniping posts. The whole unit cohesion was high. That proved to be a big help to us. We became friends with all the guys, and they would react quickly to help us, or vice versa, if needed."

The Importance of Being Professional

"What about the image of snipers?" asked Plaster. "Do others soldiers sometimes see snipers as prima donnas?"

"I don't know where that image comes from," Furlong replied. "It is not accurate. All the snipers I've met in the military or law enforcement are among the most organized, humble, and professional people I know. They are not insubordinate and always in trouble, despite the stereotype. If you don't act professional, you won't be treated professionally."

Speaking as a sniping instructor, Gilliland said, "I train my guys that to be professional, they have to run harder and faster than all the other soldiers. They have to do PT longer and harder. Their uniforms have to be better. They have to stand at parade rest to talk to NCOs or at attention to talk to officers. I have to humble the guys down, to make them real soldiers, before they can become good snipers. I tell them, 'When you walk into the front gate, you'll look like every other soldier. You do whatever you have to do when you're on an operation to get the job done, but when you're back with the rest of the soldiers, you're one of them.' It takes an incredible amount of discipline to be a sniper.

"For example, in Iraq I was an E6 section sergeant, running the sniper section, but I was also a regular grunt. I went to the tower at 3:00 in the morning to pull guard duty, and I also did KP, just like everyone else, and I made my team leaders do the same. When other people saw the sniper guys doing their part, they came to see us as 'pretty good dudes.'

"That opened so many doors for us. Things that we would never even propose at the beginning of our rotation, halfway into it were nothing. If we wanted to fast rope from the space shuttle, we could do it. If it was possible, we could ask for it, and they would do everything they could to see that it happened."

"That's an incredible change from the 1960s and '70s," Plaster interjected. "Snipers then had to sell themselves to regular units. Unit commanders and operations planners would plan a whole operation and someone would say, 'Well, what are you going to do with the snipers?' And they would say, 'We will put them on the flank, or we will use them for security at the command post.' The Army and Marines did not integrate them within the plan because they did not know how to use them. I think that is dramatically different today."

Summing up the discussion on professionalism, Plaster observed, "I have been around all three of you enough to know that you are all very intelligent, laid back, thoughtful, serious about your professional skills. None of you is boastful."

Reichert nodded. "Sometimes people will come up to me and ask, 'How many people have you killed?' You walk away. That's not what the job is about. It's not about boasting."

Proper Placement of Rounds

Turning the discussion to rounds, Plaster addressed Gilliland: "You used a 5.56mm suppressed rifle a lot, didn't you? It almost became your favorite weapon."

"It did," Gilliland agreed. "We used a 5.56 M4 a lot. I used ACOG sights, a suppressor, two-stage trigger, and Black Hills 77-grain M262 ammo. It was a very, very lethal and effective cartridge. We were taking 750-yard, first-round incapacitation shots with it. It was pretty devastating, as the terminal ballistics of that round were very impressive. It rivaled, in some cases, the .308."

"The 5.56 is very effective," Furlong agreed. "Any round is effective if you place it properly. If you put that round where you want it, it will work."

"Proper placement is key," emphasized Gilliland. "You can shoot the enemy in the shoulder or leg to incapacitate him. You don't have to wait for the perfect shot, to the head or center mass. You take the shot you can get to immobilize the body, to take the shooter out of the action, not necessarily to kill him. I'm not saying that we chose shoulder or leg shots in Iraq, but sometimes that's all you have, if someone is behind a corner, for example."

"I applaud you for that," Plaster said. "Too many people wait for the perfect shot."

"The first Iraqi I shot was in the knee," laughed Gilliland. "I took a lot of goading over that. If you can imagine: the battalion's senior sniper in Iraq, and the first Iraqi I shot was in the knee!"

"I have a similar story," said Plaster. "I shot an NVA in the foot and became known as a 'foot-shooter' after that. But he stopped shooting. That's all that counts."

Keeping a Watchful Eye

Sniping is not just about shooting, as we were reminded when the discussion turned to the role of observation and intelligence gathering in sniping.

"Your job as a sniper is to remain alert, to overwatch the guys below," Furlong told us. "You are constantly scanning, watching for things. In Afghanistan, there was so much territory to cover, your eyes would go funny looking through the scope. We used spotting scopes a ton over there. We'd look down at the ground for a second and then back to the scope to refocus. You can't be effective if you are not alert, watching."

Reichert added, "Especially with the war winding down in Iraq, they wanted to keep a lower footprint, but the commanders still wanted to know what was going on. If they didn't have an organic recon element attached to them, the snipers in the battalion were the eyes and ears of the unit."

"The best thing sniper teams are for is to confirm or deny intelligence," offered Gilliland. "When an S2 shop would come down and say to the battalion commander, 'This is what is going on in your town,' a lot of times he would come to me and ask, 'Are you seeing that?' We would confirm or deny the intel report that the S2 came up with.

"Any of the company commanders who were going out on an operation would come to us and ask, 'Hey, what's going on right here? What cars do we need to look for? What people and houses do we need to pay attention to?' That was huge for us to have company commanders we knew and trusted come to us. It completely validated everything we were trying to do.

"Hell, the Counter IED Joint Chiefs Task Force selected my sniper section in 2005 as the most successful sniper section in Iraq for our counter-IED effectiveness. After we came in, IEDs in our section of Ramadi decreased from approximately 90–100 a week to the low teens, if that, over a three-week period. It only took us going in, being at the right place at the right time, being completely disciplined, minding our manners, and engaging our targets of opportunity.

"After the first month, the IED bad guys started getting cell phone calls from the insurgency, saying, 'Hey, I need you to go and put an IED out on this or that street.' And the bad guy would say, 'No.' This is an actual conversation. 'What do you mean, no? Who are you to tell me no?' And the IED guy would say, 'If I go there, the sniper will shoot me!'"

Are Snipers Born or Trained?

The one question that every sniper gets asked (other than how many kills he has, which as we have already learned is none of your business) is whether snipers are born or trained.

"I think that being a sniper is a born trait," Gilliland offered. "It is not necessarily that you are born with the skills, but that you pick them up through your experiences in life. It's not the training you get in the military or law enforcement as much as the skills you develop during your lifetime. That doesn't mean that people from urban areas can't be good snipers."

"Shooting is a very small part of sniping," Furlong reiterated. "The mindset is the important thing. Any instructor can teach someone to be a good shooter—not a great shooter, but a good shooter. You need the right guy to train to be a sniper. You can tell who has the right skills to be a great shooter, a great stalker."

Gilliland declared, "I want the kid who fits the personality. I can teach that kid anything he wants to know. But he has to be the right kid to begin with."

One common perception is that most successful snipers hunted as kids. Plaster asked these three whether they had. It turned out that two of the three had started hunting at an early age, and hunted as often as they could.

"I hunted from the time I could walk," Gilliland said. "I lived in northern Alabama, at the foothills of the Appalachian Mountains, adjacent to a national forest. I'd come home from school, check in with Mom, and take off to hunt something."

"I grew up on the outskirts of Boston hunting anything that moved, turkey, squirrels, whatever," Reichert added.

"Me too," Plaster said. "I'd go hitchhiking with my .22, and people would pick you up and ask where you were going squirrel hunting. No one called the cops. Nowadays they'd call the SWAT team."

Reichert joked, "Now you'd be the terrorist."

Though he had grown up in a small town of about six hundred people in Canada, Furlong did not hunt as a young man.

That prompted Plaster to comment on the prevalence of snipers from small towns.

"I go down to Fort Bragg quite a bit in conjunction with sniping matters, and I notice that many of the snipers there are from Small Town, America."

Gilliland noted, "These people want to shoot, want to become snipers. You have to be a certain type of person to want to do this job. People from small towns want to become snipers. Part of the weeding process is to figure out who wants to become a sniper because it's cool and who wants to do it because it's a job at which he can succeed."

Comic Relief

Anyone who has been in combat knows that some of worst things happen in the field, but also some of the funniest. The snipers all agreed that humor plays a vital role for soldiers, including snipers.

"You need humor for mental health, for comic relief in stressful situations," Furlong said.

Gilliland concurred. "One time Harry and I had just engaged an enemy in Iraq. We were in this tiny building, with a single light-bulb hanging down. Harry, who came from a law enforcement back-ground (he was National Guard), reached up and turned on the light. I turned to him, and whispered, 'Harry, there are two people out there who want to kill us!'

"'Oh,' Harry said sheepishly, and turned off the light. When we both got back, it was hilarious. Harry was a great guy, don't get me wrong, but he came from a law enforcement background, and in law enforcement when you engage someone, it's all over."

Reichert remembered a similar occasion in Iraq. "We were pinned down on a rooftop in Iraq, doing everything we shouldn't be doing, sky-lining ourselves, watching the sun setting. Shots were whizzing over our heads. Finally, it dawned on me, and I said, 'I think someone's shooting at us!'

"'I think so, too,' someone else said. Then we started laughing. It was hilarious."

Plaster had his own story from Vietnam to share.

"Many years ago in SOG I had a friend with a recon team on a trail watch mission—count heads and see if there were porters hauling sup-plies, whatever—along the Ho Chi Minh trail in Laos. They were lying beside the trail when a long column of North Vietnamese came by. One of the NVA wanders over to the side to take a leak. The NVA looks over and he could see the Americans lying there, but he doesn't say a word. He carefully buttons his fly and leaves. One of the Americans on the ground turned to another and whispered: 'Must be a draftee.'"

It's a Wrap

The final day was dedicated to final scenes, final thoughts, and last explanations. By midday, we were ready to wrap up. Peder Lund observed that the shoot had gone incredibly well because none of the shooters had insinuated their egos into the production. "You are true professionals," he told them.

"There is one more scene to be filmed," Plaster reminded us. As the cameras rolled for the last time during the shoot in Wisconsin, he added the epilogue to *Ultimate Sniper III*:

"This sniper training film is dedicated to two sniper-trained Medal of Honor recipients, U.S. Army Special Forces Staff Sergeant Robert J. Miller and U.S. Navy SEAL Lieutenant Michael J. Murphy. Both fought above and beyond the call of duty and gave their lives in Afghanistan, defending all of us."

6

WAR IN AFGHANISTAN

Taliban Big Shot Taken Out

By Harold C. Hutchison

Not Far Enough

A Taliban big wig found out the hard way that nearly two kilometers away was not far enough when it came to being safe from a British sniper, the *London Daily Mail* reported. Corporal Christopher Reynolds of the Black Watch Regiment made the long-range shot after a three-day wait on the roof of a shop somewhere in the town of Bababji in Helmand Province, Afghanistan.

Reynolds's shot came in at 1,853 meters (2,026.5 yards), and was fired at "Mula," a Taliban commander notorious for planning a number of assaults on American and British forces since Operation Enduring Freedom began in 2001. That planning came to an end in August 2009.

British Royal Marine Sgt. Gareth Beardshaw gives distance readings to Color Sgt. Jason Walker as the sniper team returns enemy fire in Lakari Bazaar, Afghanistan, July 19, 2009. Afghan National Army soldiers and U.S. Marines with 1st Platoon, 2nd Battalion, 8th Marine Regiment worked together to deny freedom of movement to the country's enemies. The Marine battalion is the ground combat element of Regimental Combat Team 3, 2nd Marine Expeditionary Brigade-Afghanistan.
(U.S. Marine Corps photo by Gunnery Sgt. James A. Burks)

"Very Heavy Fighting"

Reynolds told the *Daily Mail* about the operation leading to the long-range shot.

"We were in a bazaar for days in some very heavy fighting and had taken up a position on a shop roof to observe the surrounding area.

"From the first few minutes after we landed, we came into contact with the enemy. We were taking fire all the time. We were observing down the valley and I saw a group of five Taliban."

Fatal Use of a Radio

In *The Ultimate Sniper,* Maj. John L. Plaster, USAR (Ret.), states that the first priority after a sniper's immediate safety is to target enemy leadership. Decapitating an enemy can have a number of effects on an opposing force. One of those cues to identifying a leader is to locate someone near a radio; another is to note whom the enemy takes direction from.

In the August 2009 engagement at Bababji, Corporal Reynolds used his training to find the Taliban commander. It started when he caught sight of a man on a radio. "I identified one straight away as the commander because I watched him through the scope and when he spoke on the radio, the other one would do what he said," he told the *Daily Mail*.

Working with Lance Corporal David Hatton, Reynolds made the calculations to ensure the shot would be on target . . . but it would involve a little trial and error.

First Shot Misses

Reynolds described the engagement to the *Daily Mail*: "We did all the calculations for range, wind speed, and all that.

"I have to admit the first round landed next to him. We were so far away that he didn't even realise he was being shot at. We changed our aim and when I took into account different factors like the trajectory of the bullet, my gun scope was actually aiming at the top of a doorway."

Second Time's the Charm

Reynolds was dead on target the second time around. This time, the "lead sleeping tablet" would hit its intended target.

"I fired and the bullet went off, coming down and hitting him in the chest," Reynolds said. The Taliban commander went down, falling into the arms of a subordinate.

"The guy just panicked and dropped the leader and ran away."

Touching off a Firefight—and a lot of Talk

Reynolds's shot was the start of an engagement. The Taliban reacted —and were backed up by a sniper of their own. However, the Taliban sniper was no match for a FGM-148 Javelin anti tank missile fired by another British soldier. It also led to a fair bit of bragging on the part of Reynolds.

"He did a top job that day—but we are all sick about him going on about it and telling us what a great shot he is," said Lance Corporal Hatton.

Sniper in Afghanistan

By Joseph DeBergalis

A brief description of the events of January 2, 2008, in the Hindu Kush Mountains of Kunar Province, Afghanistan, in the words of Sergeant Nicholas M. Ranstad, U.S. Army. At the time, he was assigned as a sniper to Recon Platoon "Hatchet," Task Force Saber, 1-91st Cavalry, 173rd Airborne Brigade.

Four Possible Acm, 2 Klicks Out

"My platoon was manning Checkpoint Delta, a traffic control point (TCP) preventing supplies going north to the anti-Coalition militia

Sgt. Nicholas Ranstad in Afghanistan with a Barrett M107 .50-caliber sniper rifle, the system used in making a 2,100-meter shot. The Barrett's magazine holds ten rounds.

(ACM) on the border with Pakistan. I had just come off of guard duty around midday and was about twenty minutes into my nap when SPC Simpson (my spotter) came running into our hooch. He informed me that SGT Lovett, who had relieved me, had just spotted four possible ACM about 2 kilometers (klicks) out. I jumped out of my cot, threw on my boots, grabbed my Barrett and equipment bag, and ran toward the guard post. As I approached, I hollered at SGT Lovett, asking him what we had. His reply was that he had four personnel, unknown if they were armed, but it was obvious that they were reconnoitering and they didn't know or realize that he had spotted them."

Dialed In

"SPC Simpson and I took a position on top of the Afghan National Police (ANP) hooch. Fellow snipers SPC White and SPC Schuch joined us to provide suppressive fire if needed and we all started to set up. All the while I could hear my platoon sergeant, SFC Condra, and SSG Morrow on the radios squawking to the tactical operations center (TOC). As I got set up and looked through my scope, I spotted the four possible ACM. SPC Simpson had my data on previous engagements (DOPE) book and dialed me in. We knew the area very well and used their exact location to gather DOPE all the time. Our initial estimate showed the ACM were 2,100 meters down the lock, a river valley between two of the mountains that marked the Afghan–Pakistani border, a reference point known as a 60-inch shot group. Our Long Range Advanced Scout Surveillance System (LRAS) truck verified the distance and we waited. As I observed the four unknowns, I confirmed that they were all carrying AK-47s and yelled back to SFC Condra and SSG Morrow that they were armed. SFC Condra relayed that information up to TOC and while he was doing that, I targeted the first one and continued to observe him. As we waited, SPC Simpson and I also established DOPE on our secondary target. Seconds later SFC Condra gave the order for weapon release. I started my breathing and [trigger] pull. SPC Simpson said 'send it.'"

Missed Low

"I pulled short, missing the initial target low, and immediately switched to the secondary as they began to scatter, not knowing where the shot had come from. The secondary hid in front of a boulder, not realizing that he had exposed himself to me. SPC Simpson, with no calculations, did an on-the-spot correction through his spotting scope. Without taking my eye from the scope or off the target, I adjusted my mildots to 7 up and 3 left and fired. Four seconds later, plus or minus, the second target began rolling down the mountain. Then, even though their targets were pretty much out of range, the others opened up, trying to provide suppressive fire. SGT Lovett was firing the M240 from his gun truck, with SPC White and SPC Schuch adding to that with the M14 and M110 Semi-Automatic Sniper System (SASS), respectively. The remaining three ACM then took cover together behind a massive boulder. I knew we had an Apache attack helicopter (call sign 'Gunmetal') on station, so I kept firing at the boulder to force their heads down and keep them from running."

ACM Waxed by an Apache

"When the Apache responded and arrived overhead, the pilot radioed in requesting target locations. SFC Condra informed the pilot that I was firing armor piercing incendiary (API) rounds at the boulder and they were deflecting off of it (the API rounds were creating a lot of sparks on impact). The pilot then radioed back that she could see the rounds hitting the boulder. She swung around, targeted the enemy, rolled in and went black on ammo, firing all of her 30mm cannon rounds and Hellfire missiles. The end result was four dead bad guys. SPC Hill (our grill man) did up a bunch of steaks on our homemade 55-gallon drum BBQ that evening. All in all, a productive day."

Hunt for the Sangin Sniper

Will a Taliban Super Sniper Strike Again?

By Harold C. Hutchison

A Sniper Killing Spree

In the town of Sangin in Helmand Province, Afghanistan, a Taliban sniper or snipers have targeted the British 3rd Battalion, the Rifles, killing at least seven British soldiers in a five-month killing spree, according to a report in the *London Daily Mirror*. One of those soldiers was hit "between the eyes." Sharpshooters from the famous Special Air Service (SAS) and the shadowy Special Reconnaissance Regiment (SRR) established in 2004 as a special forces unit have reportedly been sent to assist the 3rd Battalion of the Rifles to hunt the sniper down.

Twenty-five British snipers operate from Forward Operating Base Jackson overlooking Sangin. They deploy in teams of two, shooter and spotter, in search of Taliban insurgents.

Targeting His Counterparts

In the *Ultimate Sniper*, retired Major John L. Plaster wrote that a sniper's top priority is his own safety—and that is best protected

British soldiers from B Flight, 27 Squadron, Royal Air Force Regiment conduct a dismounted patrol near Kandahar Airfield, Afghanistan.
(U.S. Air Force photo by Tech. Sgt. Efren Lopez)

British soldiers from B Flight, 27 Squadron, Royal Air Force Regiment drive through heavy dust near Kandahar Airfield, Afghanistan.
(U.S. Air Force photo by Tech. Sgt. Efren Lopez)

by taking out that sniper's counterpart on the other side as quickly as possible. The *London Times* reports that one such sniper hunting down the Sangin sniper was killed.

"Our snipers are some of the best-trained and capable soldiers we have. When you lose one, it is telling you something," an unnamed senior British officer told the *London Daily Mail*.

But because they may not blend in like the local sniper and do not have a feel for the mountainous terrain and possible hideouts, they are at a disadvantage. Chris Hughes writes in the *Daily Mirror* of how one British sniper, Lance Corporal Teddy Reucker of 1 Royal Anglian with nineteen Taliban kills, was dispatched to chase down one illusive sniper who would shoot, then disappear in Kajaki, Helmand Province in 2007. Teddy and his company found that the Taliban sniper operated out of "killing hole" tunnels dug in thick walls. The sniper, with seasoned survival skills for the harsh Afghanistan theater, killed a snake with a knife as he was hiding out in one of the holes. He fired with a damp blanket over his rifle, hiding "muzzle flash" from deep inside the buildings.

He was believed to be a veteran of the 1980s Afghan war or the fighting in Chechnya. He was taken out by an air strike after an extensive one-month man hunt.

The Hunt for the Sangin Sniper

British soldiers speculate that the invisible sniper of Sangin must be stalking his targets for days and that the sniper was most likely trained with the help of Al Qaida in Iran or Pakistan. "The conclusion is the Taliban have outside help from either Iran or al-Qaeda in Pakistan," another officer told the *Daily Mirror*.

Where did this Taliban sniper come from and who is providing his weapons?

In 2007, the *Washington Times* reported that Iranians were arming the Taliban insurgency, supplying the insurgents with Iranian weapons used by insurgents in Iraq, particularly explosive-formed projectiles that were used in improvised explosive devices. In October 2009, the *Marine Corps Times* reported that American military

U.S. Marines with Charlie Company, 1st Battalion, 3rd Marine Regiment; U.S. Soldiers with the 630th Engineer Company; British soldiers with Alpha Squadron, Cavalry Regiment; and Dutch soldiers meet near Marjah, Afghanistan, February 21, 2010. Their units were part of Task Force Helmand, whose members were creating a plan to improve security in the area. (U.S. Marine Corps photo by Cpl. Albert F. Hunt)

personnel believed that Iranian Quds Force personnel were active in the Afghanistan–Pakistan region and were "probably" helping to kill American troops.

Public Support Falling

The sniper's campaign has come when public support for the British deployment in Afghanistan has been falling. In July 2009, prior to the start of the latest sniper attack spree, the Politico reported that the number of British troops killed in Afghanistan had exceeded the number killed in Iraq. Prime Minister Gordon Brown has come under fire for the losses, many of which were in Sangin.

"The casualty toll in Sangin is tragically high but our forces remain very much on the front foot and are determined to maintain the progress that they and their predecessors have achieved," Major General Gordon Messenger told the *Times of London*. Fifty-three servicemen have been killed around Sangin in the last year, twelve times the average death rate for NATO forces in Afghanistan, the *Daily Mail* reported.

The manhunt for the sniper continues.

SEAL Sniper Sees Action in the 'Stan

By SEAL Chief Petty Officer Brandon Webb, USN (Ret.)

Patch designed by Brandon Webb

Innocent Farmer or Taliban Sentry?

Friday, January 11

The next day we went out on another village op. Genuine bad guys were hiding out there, and people were just living up here in the mountains, usually with a wife and couple of kids, living the simple life, farmers wresting their keep from the land. We could usually tell the difference pretty clearly, but not always. There was a place we'd been watching for a few days now. These people appeared to be farmers, but we were not 100 percent positive. We decided it was time to

Navy SEAL sniper Brandon Webb with friendly indigenous forces.

go out there and see up close. I was set up as sniper overwatch to guard the platoon as they went in to meet the people and talk. It was morning. I watched as Cassidy and his team made their way up to where a small group of these guys was congregated in a few buildings. It wasn't like we were storming the place; this was more of a diplomatic mission.

Having dug into my sniper overwatch position, in the kind of well-concealed hide we'd been trained to construct in the stalking phase of sniper school, I used the scope on my .300 Win Mag sniper rifle to get a closer look at these people. The villagers clearly saw Cassidy and the guys approaching. Something was going on there, but I couldn't tell what. They were talking something over, looking a little hurried about it. I caught a glimpse of a few of them running around, as if they were in a rush to get something done. Something felt suspicious about it to me, but it could also be completely innocent. I relayed my observations to Cassidy on the radio and told him to be on his toes.

As I continued moving my rifle in a small oscillating arc, shifting my view back and forth between Cassidy and his team and the little knot of Afghan farmers, I noticed one guy standing off to the side. He had a gun.

Shit.

The man had his rifle slung casually over his shoulder, and there was nothing threatening about the posture. I couldn't tell if he was a bad actor or an innocent farmer. I was leaning toward farmer, but why was he carrying a gun? Alarm bells were going off in my head.

Cassidy and the team were now close to the house. *Man oh man,* I was thinking, *do I take the shot? Will it put Cassidy in a tough spot?*

The guy was about 600 yards away, slightly more than six football fields. I knew I could take him out in a heartbeat. No problem. I felt my finger against the trigger. Breathe out . . . focus . . . squeeze . . . pop. It would be that easy.

But if I did, it would certainly complicate the situation. If I shot the guy and it turned out he was innocent, we'd have quite a scene on our hands. However, if I didn't and he wasn't innocent, the team could be in danger. Even if these guys had more arms stashed close at hand, Cassidy and our guys would clearly outgun them. But you don't want to let things get so far that the question of who outguns whom is your determining factor.

Shit!

What do I do? I had all the information I was going to have. There was no more intel to weigh, no path of logic to make the wiser choice. It came down to pure instinct. Do I take the shot, or not?

I breathed out . . . focused . . . squeezed . . .

I decided not to take the shot.

A moment later Cassidy and the guys were there, talking to these Afghan farmers, and suddenly I caught a glimpse of movement way off to my left. Some character in Arab dress, clearly not Afghan, was hightailing it out of there, tearing along a little goat trail up the mountain toward Pakistan for all he was worth.

Motherf......!

SEAL Team Three ECHO Assault Element, Post Zhawar Kili, Bagram, Afghanistan.

Erring on the Side of Caution

This guy could have been out there on his own, but I didn't think so. They'd been hiding him. That's what I'd been sensing. The Afghan farmer I'd been targeting had been standing sentry, trying very hard not to look like that was what he was doing. They were covering for this al Qaeda dude or whoever he was, and the moment they had Cassidy and his team engaged in conversation, one of them had told him to take off. I switched to my binos and caught him scurrying up the mountain, closing in on a kilometer away. I couldn't get an accurate shot off in time, and I couldn't go after him, because to do that I'd have to leave my hiding spot and would no longer be supporting Cassidy and the team. I didn't have the radio resources to call in close air support, and in moments that son of a bitch would be over the border.

I got back to Cassidy on the radio and told him what had happened. I could see him now, going back and forth with the farmers, who were hotly denying everything. I'd seen enough to know they were lying. Thinking back over the whole sequence, I didn't see what I would have done differently. With the information I had, giving this farmer the benefit of the doubt still seemed to me the right decision. Yes, these Afghan village people would sometimes harbor other Afghans who were Taliban or Arabs we would call al Qaeda. For the most part, though, they were not bad people; they were just trying to get along and survive, to go on living there in the mountains the way they had been for generations without getting caught in the crosshairs of battle.

Deployed During the 9/11 Hysteria

When we first arrived, in Kuwait and Oman and finally Afghanistan, we were hyped up and angry and ready to deliver payback. We were coming right off the shock of 9/11, and we had all sorts of people emailing us from the States, voicing their support and cheering us on. Underneath that caricature of the white devil and "3 ECHO" on our platoon patch, I'd had a legend stitched that said, EMBRACE THE HATE. That's the mode we were operating in, and our rules of

Navy SEAL Brandon Webb with German KSK troops at remote northern Afghan checkpoint.

engagement certainly supported that. When in doubt, take them out. However, as we got more immersed in the culture and started seeing things from the point of view of the people who lived there, things began to shift a little. I'd been in Afghanistan long enough now to understand that not everyone had to die. I didn't want to shoot anybody who didn't need shooting.

Still, the shot I didn't take sometimes haunts me as much as some of the shots I did.

A Two-Man Mobile Surveillance Unit

Saturday, January 12

By Day Seven, we were starting to wrap up the operation and prepare to return to base. We had now been holed up in this mountain range for a week and had cleared out a ton of enemy resources, taken a handful of prisoners, and racked up dozens of enemy killed in action (KIA), but there were still a lot of bad actors in the area that we hadn't been able to track down. Even our surveillance tactics of a few days ago had had limited success. Sitting in that one spot for the whole day, we weren't able to observe nearly as much as we'd have liked.

Osman and I had an idea. We wanted to get out there on our own, just the two of us, and patrol the area without having to be tied to a whole squad: a two-man mobile surveillance unit.

We pitched the idea to Cassidy. We proposed that the two of us go out, insert at two in the morning, and spend the entire day scouting the area. See what was really going on out there and what we could turn up.

There was a checkpoint we had observed, maybe five miles south of our position, a controlled-vehicle access point usually manned by two to four guys at any one time. Because of the Army Special Forces incident that had mistakenly taken out a bunch of Karzai's people, we were especially cautious about making sure who these guys were before we took any action. Osman and I had been watching these guys for days, and by now we were clear that they were Taliban. They were facilitators, ground warriors whose primary mission was to run combat supplies back and forth across the Afghanistan–Pakistan border—money, passports, intel, and other tools of the trade. We also knew they'd been surveilling our own platoon. Hell,

SEAL Team Three ECHO Assault Element, Task Force KABAR, Kandahar, Afghanistan.

they'd nearly ambushed us more than once. Now we wanted to go out counter-surveilling the guys who were surveilling us.

Cassidy and Chief Dye gave us the thumbs-up. That night, we mapped out our route, got our plan together, and packed up our kit. We headed out early the next morning, about 0300. Before leaving, we had checked in with Mark, who was our comms guy, and told him what we'd be doing throughout the day and to make sure to check us in with tactical operations center (TOC) and let that C-130 know they'd have two friendlies out there. We had our infrared (IR) glint tape on, a special reflective tape like joggers wear at night, except that instead of reflecting visible-range light, it reflects infrared. We hoped they'd see that, but it sure wasn't something we'd want to count on. We did not want to be little green heat trails in the C-130's video game.

Stalking in Traditional Afghan Gear Camouflage

By this point, Osman and I were totally garbed out in traditional Afghan gear. We were wearing wool shawls and Afghan roll-up hats; we had water, bullets, a little food, and guns. At a casual glance, we could have passed for Taliban. We'd gotten the lay of the land and were now running around those goat trails, too. As much as it was possible to do, we had become mirror images of the guys we were about to hunt.

We got to the bottom of the hill, humped over to our first observation post, settled in, and waited for the light to come up. Osman looked over at me and said, "Sure hope Mark called that damn C-130." I nodded. I sure hoped so, too. We were pretty vulnerable out there and had put our lives in Mark's hands.

The cold morning air hung thick in the valley. Each warm exhale of breath briefly fogged the outside corner of my scope as I waited and watched.

There.

I could just make him out: a middle-aged man, wrapped in traditional Afghan dress, darting furtively back and forth and breaking down his makeshift campsite with seasoned efficiency. I noticed a slight crook in his step—an old wound, perhaps a story from the days of the Soviet occupation.

The man had been at his clandestine trade for years. He would be at it for less than twenty-four hours more. I saw a faint wisp of smoke from the campfire he had just extinguished, and my brain automatically registered the direction and intensity of the gust of breeze that flirted with the smoke, calculating windage, distance, and elevation. We could take him out right then and there; Cassidy had given us the go-ahead. If we did, though, it would likely be our only kill of the day, because the moment you fire your weapon, you've risked compromising your position, and you never know who else was lurking around the corner or somewhere behind you, especially in an environment like this. Besides, we had a bigger strategic goal. We could kill one, or we could find them all, mark their positions, and they would all die.

I've since been deer hunting quite a few times; that's what this was like, except that we didn't expect to shoot anyone. Today it was not our marksmanship we'd be practicing, but our stalking craft. As much time, energy, training, and focus as we put into our marksmanship skills, the core skill of the expert sniper is not to shoot. It is to hunt. If intellectual capacity is a sniper's foremost qualification, the number two trait is patience. We will take out any enemy we have to when the situation calls for it, whether that means using a rifle, a handgun, a knife, or our bare hands. Yet the sniper's fundamental craft is not killing a person, but being able to get close enough to do so. Osman and I were on a classic sniper stalking mission: track, sneak up, observe, and disappear again, leaving no trace behind.

A Gigantic Trap

The man was moving out now, ready to start his day. So were we. A short while later we found the spot. The man and a few of his cohorts had been using this site to lay up at night: bedroll stash, food and water, some ammo, evidence of a small fire for cooking. Chances were very good they'd be back that night. We marked the GPS coordinates and backed out again, leaving everything exactly as we found it, and moved on.

We spent the day out there, covered a good 10 to 12 kilometers and located about half a dozen sites. We got back to camp about midnight. After reporting in, we sat down and put our notes together, lining up all the coordinates so we had a tight sequence. By this time, we were already familiar with the process of calling these coordinates in ourselves. Brad and Eric had spent so much time over the week calling in air strikes that they'd gotten some of us to spell them at times, just so they could take a break to eat and get some rest. By this point, we had already called in a lot of ordnance in this valley.

Now, in the middle of the night, they set Osman and me up on the radio, and we called in our sequence ourselves. We had laid a gigantic trap, and now we would be the ones to spring it.

The site we had occupied with Chief Dye that first night at Zhawar Kili gave us an amazingly clear view of the valley below, such that we were able to gaze out with our binos and get an easy visual on all the locations we'd marked during the day. One by one, we saw the barest flicker here, a glint there, telltale flashes as they fired up their cookstoves and campfires signaling us that, *yes, this site was occupied again tonight*. We called in our coordinates, one by one.

Boom! Boom! Boom!

One after the other, we called in the numbers to our F 18s overhead and sent them all to hell.

A Fort Knox of War-Making Wealth

Sunday, January 13

After a week of forensic spelunking, and even with all the air strikes we had called in, we knew we still had not come close to destroying all the equipment, weaponry, living supplies, and other material that was stashed away in that mountainside. This place was a Fort Knox of war-making wealth. There was no way we could carry all this stuff out with us, and we didn't want these guys coming in here after we left and digging out their stashes of ammo and whatever else they might be able to find. So we choreographed one last hurrah. All the intel we'd gathered over the week was orchestrated into one final bombing session, the largest since the bombing of nearby Tora Bora exactly one month earlier. We pounded that place, and caved in the side of the mountain.

Our twelve-hour mission had turned into a military and political bonanza. In a network of more than seventy caves and tunnels, we'd uncovered nearly a million pounds of ammunition and equipment, along with a ton of intelligence, including extensive papers documenting cross-border traffic and other aspects of enemy tactical plans. More than 400,000 pounds of ordnance was dropped on the targets we flagged. We had destroyed one of the largest terrorist/military training facilities in the country and had taken out a significant number of enemy personnel.

The following day, Monday, January 14, on the ninth day of our twelve-hour mission, we boarded a pair of helos and lifted out of Zhawar Kili, bound for Bagram and Kandahar.

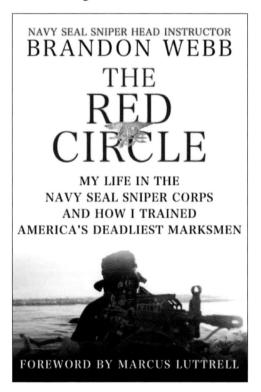

NAVY SEAL SNIPER HEAD INSTRUCTOR
BRANDON WEBB
THE
RED
CIRCLE
MY LIFE IN THE
NAVY SEAL SNIPER CORPS
AND HOW I TRAINED
AMERICA'S DEADLIEST MARKSMEN

FOREWORD BY MARCUS LUTTRELL

Deadly Battle in Sangin

By Sgt. Tyler Hughes

I looked back at the Osprey, its propeller blades lumbering to start, as they began to beat faster and faster, kicking up dirt, mud, and the remains of freshly planted crops. Before long, the rotor wash had showered us with the farmland we were standing in. We watched as the Osprey ascended into the night sky, leaving us behind enemy lines in Taliban country. This was a place where we had no friends, only enemies. Once the roar of the Osprey had faded, we looked at each other; I and the four other men that comprised my Scout Sniper team gave one another a nod and started to move. The soggy soil of the farmland slowly engulfed our legs as high as our thighs. It sloshed beneath us, providing a rhythmic sound in the utter stillness of our surroundings. It tried to slow us down and make the first leg of our three-leg journey miserable; it was warning us of things to come.

After 500 meters, we reached the canal that would comprise our second leg. We quickly climbed into the water and it swallowed us to our belly buttons and chests; we took a moment to acclimate. I raised my hand, gave my guys the cue, and we stepped off. At this point, the 120 pounds of gear and packs each person was carrying became completely soaked, water permeating every pocket and crevice. The water slowly crept up our backs, the 120 pounds slowly starting to feel like an extension of the canal below, trying to drag us down and swallow us whole. With our measured steps, two miles felt like twenty and 90-degree heat in the middle of the night wasn't offering reprieve. But if there is one thing marines know, it's pain, and it is something that I grew to love.

The water was dripping from my fingertips and the bottom of my body armor back into the canal. I looked up at the mountain that stood two kilometers away and then looked at the men in my team. They were scanning the mountain with the same determined look on their faces. Topping out at 350 meters, this mountainside was as steep as it was rugged. There was one small road and that was the only way up the mountain. It was a narrow corridor and if we used it in the dark of the night, we would not be able to spot any of the IED indicator signs that we could in daylight. I signaled my team to find a place to hide and wait for nautical twilight; at least that way we could have enough light to guide us up this beast of a mountain.

Scaling the Steep Mountainside

As nautical twilight came, we started to traverse the mountainside. It was so steep at times we had to rope each other up, all while being cognizant of the threat of IEDs and the even bigger threat of being spotted and ambushed. The sun was relentless as the temperature reached triple digits. It felt like the terrain and the weather were conspiring to make this trek impossible. But that is not a word in our vocabulary. We came upon a corridor where the road was carved into the mountain, narrow and long; the mountain formed walls on either side of us with no vantage point above. As I regrouped my team, we started the patrol up this path. When we'd hear rustling, we felt an

immediate sense that an ambush could rain down on our heads at any moment. Each step was cautious, but quick. We emerged from that path and continued to near the top of the mountain after two hours of climbing. At the very top of the corridor, we found two little packages waiting for us in the form of IEDs. As we carefully stepped around them, I gave them a silent "F. YOU," hoping that the heavens would deliver my message to the coward who had placed them there, relishing in the thought of destruction and blood it would create. But we had a mission to get to.

When we reached our destination, we established our hide site. From this mountaintop, we would live for three days and provide constant overwatch support to provide real-time combat reporting and long-range precision fire on enemy combatants. We were to be scouts and snipers. We were supporting a company level operation, whose mission was to start in southern Sangin and over the three days push north, clearing the area of insurgents. Our role was crucial. Sangin was considered one of the bloodiest places in the world for American forces for a reason. Unlike my previous deployment to Afghanistan, where villagers were more accepting of coalition forces, this was a place that was firmly on the Taliban's side. They didn't want clean water, they wanted blood. Our blood. Our battalion was consistently sustaining substantial losses, both in KIAs and injuries. The enemy was able to hurt us with IEDs and bullets; but even in combat operations, there were times that they were one step ahead of us. They knew where squads were heading and what they were doing. While the battalion was able to monitor radio traffic and gain valuable intelligence, there were instances when the enemy was able to utilize other forms of communication to their advantage.

An Enemy Spotter

On day three of our mission, we were providing over-watch when a squad started their operations into a village in the northern end of Sangin. They were taking fire, returning fire, and suppressing the enemy. As the squads changed course and changed plans, all while being extremely careful to not reveal themselves, they continued to take fire. Suddenly a glimmer flashed in my scope. I wasn't sure what it was; at first it didn't seem like anything, but I couldn't ignore it.

I kept watch like a hawk with my M40A5 through my Schmidt & Bender Scope, 1,000 yards away, waiting for something.

A few minutes passed and nothing happened, and then a man appeared, cell phone in hand, in an area with tall grass. My entire team was now devoted to this 200-meter stretch of riverside. Every time the squad would do something, the man on the cell phone would pop out quickly to evaluate what was happening and then, within seconds, the squad would be taking fire again. This happened over and over and over again. It became a pattern. During this time, we noticed a second individual with the spotter, who appeared to watch his back, ensuring they weren't surprised from the rear by coalition forces. I quickly decided that the cell phone man was a spotter for the enemy, a person whose sole purpose was to provide the enemy shooters real-time information on what coalition forces are doing. They lacked weapons, because of their proximity to coalition forces, so that if they were caught they could pretend like they were poor farmers, tending to their land in the middle of a fierce firefight.

My spotter, using his range card, gave me the range and angle to the target. At this time, I radioed to an adjacent sniper team, requesting they engage as well. The other team rogered up; they were on board and preparing for the controlled shot. Ensuring that HQ was aware of our actions, we communicated with them one final time. I had the enemy spotter in sight, my crosshairs steadily on his chest. The actual distance was 960 meters. I applied the elevation adjustment to my scope, double checking that I had applied it correctly. My spotter gave me a wind estimate. I checked to ensure I was seeing what he saw—a full left-to-right wind value. I applied this wind correction to my scope. We contacted the adjacent sniper team, relaying our wind reading, with which they concurred.

My spotter, looking through his Leupold spotting scope with the radio handset to his ear, said "Team 1 will engage the actual spotter, Team 2, you will engage the lookout individual. I have control, firing on the T of Two. 5 . . . 4 . . . 3 . . . 2 . . . Bang." Two shots rang out in unison.

My shot struck the enemy spotter in the lower chest. He fell down into the grass. The adjacent sniper team missed with the first round, impacting high over the second individual. Now confused, wondering where the shot came from, this individual froze, squatting and attempting to locate a safe place to hide. The adjacent team reengaged, striking the second individual in the upper right shoulder. We continued to observe the area, searching for signs of life. Nothing. Approximately fifteen minutes later, three individuals with a wheel barrel exited a building complex 50 meters away, heading toward the two lifeless enemy combatants. This was the dead body pick up crew. They collected the two lifeless bodies, heaping them into the wheelbarrows and running off with slumps of terrorist scum to dispose of. I looked around at the team; I saw three dirty, grungy smiles. We didn't speak, we still had a mission to complete, we had to keep our composure, but it was clear we were all thinking the same thing—we had just killed two enemy combatants and saved marines' lives. What better reason was there to smile?

I have seen the best of mankind in war and I have seen the absolute, most vile side of humanity. Men who use children to do their bidding, sending them out to be killed or to befriend coalition forces to lure them into death traps. I have seen families rejoice when schools have opened and clean water has been brought to their village. I have seen people hell-bent on the destruction of men like myself, people who will stop at nothing to make sure that they bring chaos and oppression to their own. I look back at my job during this deployment, and especially during this mission, to ensure that all of my brothers stepped back onto the FOB that night. I was determined to prevent any injuries and deaths. If I could prevent it, there would not be a flag-draped coffin greeted by a grieving widow. Not while I was looking through my scope. My battalion saw nineteen men killed and hundreds wounded during our time in Sangin. Even though I cannot change those numbers as they are, I know that the men of my team and I made sure those numbers weren't higher, and it is for that reason I can sleep soundly at night.

7

WAR IN IRAQ

Crosshairs at the Crossroads

By Maj. Jim Land, USMC (Ret.)

*Maj. E.J. Land, Executive Officer Weapons
Training Battalion, Quantico, Virginia, 1970*

The fate of United States Marine Corps snipers is at a crossroads. The tragic deaths of six snipers and their security personnel over the past few months in Iraq suggest the path the Corps is taking may have grave, unintended consequences. Not the least of those consequences is a dilution of the mission, training, skills, and safety of these very unique combatants, to the degree that further tragedy in Iraq or wherever Marines find themselves in harm's way may be inevitable.

A close look at the circumstances that led to the sniper deaths in Iraq reveals a wealth of information about the state of snipers and sniping today, the bulk of which appears strangely self-contradictory. One very important outcome, however, is the bittersweet opportunity to openly and candidly discuss the nature, role, and deployment (past, present, and future) of Marine snipers. It's a discussion that needs to be held among the highest echelons of Marine Corps leadership as well as among those marines in the field from Division to the individual combatant hunched behind cover awaiting his next adrenalin-laced set of orders.

In one very real sense, the deaths of these young marines are directly attributable to the success of Marine snipers in general, and of Marine snipers operating in Iraq in specific. At no time in U.S. Marine Corps history has the role of the Marine sniper been so visible to so many. Therein lies the crux of the thesis of this article.

Shock, Awe, and Public Sniping

On the night of March 21, 2003, America and the world watched as ton after ton of precision bombs signaled the official start of Operation Iraqi Freedom. Viewers continued to hold their television vigil on April 9, when images of the first units of marines entering Baghdad cycled around the world to be fed into the nightly news. Those young marines promptly set about helping Iraqi citizens topple a massive, far-larger-than-life statue of Saddam Hussein that dominated a public square.

Arguably, our digital society's ability to ramp up the visibility of images of war broadcast to the public parallels the equally uncharacteristic visibility and unprecedented recognition the sniper is receiving from fellow marines, including commanding officers in Iraq, versus the virtual anonymity surrounding their ventures in the jungles of Vietnam.

Traditionally, Marine sniper teams operate apart from the majority of fellow combat troops. They ply their lethal trade under deep cover, purposely isolated from observing eyes, from a multitude of ever-changing positions. Operating under stealth conditions keeps the sniper alive in the field. Unfortunately, the cloak of secrecy associated with sniping also perpetuates a mix of emotions, even among fellow Marines, about the character and mission of the sniper. Some verge on myth; some can be downright hostile. That quest for anonymity also speaks volumes about the true character of the Marine Corps sniper.

A Cadre Apart

The deliberate isolation of sniper activity from other frontline troops allows the sniper to focus entirely on his mission and to employ the skills vital to that mission's success. Throughout the duration of Vietnam, no Marine sniper—including the late, legendary Carlos Hathcock—was ever awarded a commendation for feats of combat sniping marksmanship. Then, only members of sniper teams knew with absolute certainly what the man behind the trigger accomplished. During Vietnam, it was not in the Marine sniper's nature to seek approval by medal for doing his job. While the Marine sniper's training and character have not changed, the current mode of conducting warfare and the way Marine snipers are commanded and deployed in Iraq is unintentionally undermining virtually everything the sniper is taught—lessons that save Marine lives. The Iraqi conflict has already seen Marine snipers on the receiving end of two Bronze and one Silver Star for their combat rifle craft.

Even that much deserved recognition plays a major factor in the threat to those same young Marines. Snipers prefer a low-profile existence in every sense of the word. Traditionally, they seek only the inward satisfaction of a job well done. By training and, some might argue, by their genetic makeup, the Marine sniper is first and foremost the consummate team player. The sniper provides invaluable support to his fellow combatant's with each shot fired. Whether the projectile's target is selected from the opposition's command staff or its pool of grunts, certain death (at some distances, silent) instills paralyzing fear in an enemy and gives fellow marines an edge in combat.

The increased visibility of the Marine sniper among fellow marines and unit commanders associated with the present Iraqi war is one of the pivotal problems the Corps must identify and deal with if Marine snipers are to accomplish their mission and preserve their lives. Visibility flies in the face of virtually every trait of the Marine sniper.

Historically, the term "sniper" originated among British Imperial troops stationed in India during the eighteenth century. Then the marksman able to hit the speedy and seemingly impossible to shoot little birds was dubbed a "sniper." In combat, however, the effective use of the combat sniper is attributed to the German Army of World War I. The American military sniper's genealogy might also be traced to the then-unconventional fighting style of the Revolutionary War era hunter/citizen–soldier that sent well-aimed rounds from custom-crafted long rifles, from behind cover and concealment, toward the smoothbore-armed British and Hessian occupational troops.

The Enemy As Prey

The emphasis on "hunters" and "hunting" is deliberate. The skills developed in the forest and field—stalking, observation, understanding the nature and habits of the prey, avoiding detection, and delivering a single lethal shot from a hidden position—and those personal characteristics associated with hunting cultures around the world—are the same U.S. Marine Corps Scout Sniper training instills in its graduates.

The fraternity of the world's great hunting cultures is the closest analogy to the brotherhood of U.S. Marine snipers. They are experts in stalking, observing, and the art of precision long-range shooting. Like the Arctic, African, or Native American hunter, the sniper above all else observes every detail of the environment around him. Depending upon the terrain and geographic locale, the wind, trees, grass, birds, even the insects provide vital intel.

Like the ancient hunter in search of bison, elk, or antelope, the sniper holds no malice for his enemy. He develops a kindred feeling that comes from endless hours watching, studying, and learning his quarry's every move. He respects his enemy and demonstrates that respect through the discipline, conviction, and ethics with which he conducts his every action in combat.

Spoiled by Success?

Unfortunately, the early days of sniping in Iraq brought such unbridled success in dispatching large numbers of poorly trained and confused enemy personnel in record time that the snipers' sense of respect for his foe all but vanished. Young Marine snipers grew complacent in their open disrespect for the intelligence of the typical veteran of Saddam Hussein's military apparatus. That seductive air of complacency, where respect for the enemy gives way to arrogant bravado, can and does prove fatal—to marines, not the enemy.

The period where the opposition encountered by U.S. forces was mainly a display of sheer buffoonery by Saddam's ill-trained and confused military personnel, where Marine snipers could log large numbers of confirmed kills, as high as fifty to sixty, taken down at a rate, in some cases, of a dozen or so in twenty to thirty minutes, is no more. Now, highly trained and combat-seasoned professionals from Bosnia, Afghanistan, and other Muslim military training centers of anti-American hatred are mounting the bulk of the Iraqi insurgency.

That transition to a well-trained enemy is all too graphically illustrated where the loss of a U.S. Marine Corps M40 sniper rifle quickly translates into single-round, headshot fatalities among U.S. combat personnel less than twenty-four hours later.

Another vital lesson as yet not learned in Iraq regarding the best use of Marine snipers also stems from the initial success of their marksmanship prowess. As every present and former drill instructor will attest, "a little bit of knowledge" more often than not results in big trouble. That bit of DI wisdom, unfortunately, aptly describes the all-too-familiar use of snipers in Iraq today. It is causing sniper resources to be deployed in ways that intrinsically undermine the ability of the sniper to function as he has been trained.

As currently configured in Iraq, Marine snipers operate at Battalion-level command. With no disrespect intended, Marine officers unfamiliar with the tradition and full value of sniper deployment are impressed with the efficiency and skills of the Marine sniper. They see the tilling power of the sniper's M40 bolt gun and recognize the sniper's ingrained observational skills. They then fall into the "one size fits all," thinking that the skills of the sniper make him the perfect candidate for a variety of other wartime tasks. Consequently, snipers are being inserted into reconnaissance roles to gather G2 Intel.

Scouting versus Intel Gathering

This mutation of the Marine sniper is also reflected in the gear he's required to carry. Unlike the lean years of Vietnam when knife, ammo, water, a few grenades, and very little else were the only accessories carried into the bush, the Marine sniper in Iraq is expected to deploy with a full rucksack.

And while the M40 bolt-action sniper rifle is considered the premier arm for one-shot precision fire from a concealed position, the potential of running into close quarter contact trouble on a recon mission demands the added firepower of an M16 at the very least.

Worse, and this is the single most culpable reason for the unacceptable degree of sniper loss in Iraq, the first premise of sniper deployment is being countermanded by orders issued from Battalion by officers who simply do not have the first clue of proper sniper technique. Snipers are being ordered to return multiple times to previous positions to observe enemy operations.

The truth of the matter is that the tragic incidents referenced at the top of this article that saw snipers ambushed occurred as a direct result of their orders to return to posts previously abandoned. In each incident, the enemy was waiting and the hunters became the hunted. Snipers are taught to never, never establish a pattern of activity. Predictability is the way most prey—whether four-legged or two—is tracked, trapped, and lulled. That's why snipers, the successful ones, are undetectable, unpredictable, and constantly on the move.

Snipers are not prima donnas who demand special treatment and care. They are important combat resources that provide vital support for combat operations. Their job requires the complete focus of their every sense and razor sharpness of their entire mental faculties. They must be well-rested in advance of a mission if they are to bring into play the multiplicity of skills and attention to detail that successful sniping demands. That includes the ability to move undetected into position, to determine that position from their on-site evaluation of the immediate landscape, and their ability to pick and choose targets as they see them fitting into their mission.

Intel Is Secondary

Intel gathering is not their key mission. The mission of the sniper is, put in the most cold and candid terms possible, "to kill."

Their ability to achieve that objective instills fear in the enemy. The Iraq experience is proving the correctness and immediate need for the adoption of a much-discussed-and-debated concept of exactly where snipers best fit within the Corps organization table.

Specifically, the fate of U.S. Marine Corps snipers depends upon the very demonstrable need for a Scout Sniper Company at the Division level. Basing snipers at the Battalion level just doesn't work. Iraq is the on-job-training proof of that faulty pudding.

Until and unless all Marine officers go through the Corps Scout Sniper training, a highly unlikely scenario, locating a sniper company—commanded by sniper-trained officers—is the only logical solution to both the needs of the Corps and the safety and continued effectiveness of its snipers for generations of warfare to come.

Creating a sniper company operating out of the Division is not without precedent and is quite logical if snipers are thought of in a capacity similar to artillery, air, armor, and reconnaissance support, all of which are on-call for battalion needs. The U.S. Marine Corps sniper is every bit as unique a combat support weapon as Marine Air, Armor, or Artillery. Deploying the Corps sniper resources similarly is the logical solution to the tragic problems being demonstrated in Iraq today.

The Weaponry of America's Most Lethal Sniper

By SEAL Chief Petty Officer Chris Kyle, USN

A U.S. Army soldier looks through the scope of his M14 sniper rifle near Howz-e-Madad,
Kandahar province, Afghanistan, January 12, 2011. An operation was conducted to clear
the area around the village near Howz-e-Madad of insurgent forces.
(U.S. Army photo by Cpl. Robert Thaler)

I was not the best sniper in my class. In fact, I failed the practice test. That meant potentially washing out of the class.

Unlike the marines, in the field we don't work with spotters. The SEAL philosophy is, basically, if you have a fellow warrior with you, he ought to be shooting, not watching. That said, we did use spotters in training.

After I failed the test, the instructor went through everything with my spotter and me, trying to figure out where I'd gone wrong. My scope was perfect, my dope was set, nothing was mechanically wrong with the rifle.

Suddenly, he looked up at me.

"Dip?" he said, more a statement than a question.

"Oh . . ."

I hadn't put any chewing tobacco in my mouth during the test. It was the only thing I'd done differently, and it turned out to be the key. I passed the exam with flying colors—and a wad of tobacco in my cheek.

Snipers as a breed tend to be superstitious. We're like baseball players with our little rituals and must-dos. Watch a baseball game, and you'll see a batter always does the same thing as he steps to the plate—he'll make the sign of the cross, kick the dirt, wave the bat. Snipers are the same way.

During training and even afterward, I kept my guns a certain way, wore the same clothes, had everything arranged precisely the same. It's all a matter of controlling everything on my end. I know the gun is going to do its job. I need to make sure I do mine.

There's a lot more to being a SEAL sniper than shooting. As training progressed, I was taught to study the terrain and the surroundings. I learned to see things with a sniper's eye.

If I were trying to kill me, where would I set up? That roof. I could take the whole squad from there.

Once I identified those spots, I'd spend more time looking at them. I had excellent vision going into the course, but it wasn't so much seeing as learning to perceive—knowing what sort of movement should get my attention, discerning subtle shapes that can tip off a waiting ambush.

I had to practice to stay sharp. Observation is hard work. I'd go outside and just train myself to spot things in the distance. I always tried to hone my craft, even on leave. On a ranch in Texas, you see animals, birds—you learn to look in the distance and spot movement, shapes, little inconsistencies in the landscape.

For a while, it seemed like everything I did helped train me, even video games. I had a little handheld mahjongg game that a friend of mine had given us as a wedding present. I don't know if it was exactly appropriate as a wedding present—it's a handheld, one-person game—but as a training tool, it was invaluable. In mahjongg, you scan different tiles, looking for matches. I would play timed sessions against the computer, working to sharpen my observation skills.

I said it before and I'll keep saying it: I'm not the best shot in the world. There were plenty of guys better than me, even in that class. I only graduated about middle of the pack.

As it happened, the guy who was the honor man or best in our class was part of our platoon. He never had as many kills as I did, though, at least partly because he was sent to the Philippines for a few months while I spent my time in Iraq. You need skill to be a sniper, but you also need opportunity and luck.

Guns

People ask a lot about weapons, what I used as a sniper, what I learned on, what I prefer. In the field, I matched the weapon to the

U.S. Marine Corps Sgt. Mark Hammet, a sniper with Scout Sniper Platoon, Headquarters and Service Company, in support of India Company, 3rd Battalion, 5th Marine Regiment, Regimental Combat Team 2, uses his rifle combat optic during a company operation in Sangin, Afghanistan, January 21, 2011. The battalion's mission was to conduct counterinsurgency operations in partnership with the International Security Assistance Force in Afghanistan.
(U.S. Marine Corps photo by Cpl. David Hernandez)

job and the situation. At sniper school, I learned the basics of a range of weapons, so I was prepared not only to use them all, but also to choose the right one for the job.

I used four basic weapons at sniper school. Two were magazine-fed semi-automatics: the Mk 12, a 5.56 sniper rifle; and the Mk 11, a 7.62 sniper rifle. Then there was my .300 Win Mag. That was magazine-fed, but it was bolt-action. Like the other two, it was suppressed, which means that it has a device on the end of the barrel that suppresses muzzle flash and reduces the sound of the bullet as it leaves the gun, much like a muffler on a car. I also had a .50 caliber, which was not suppressed.

Let's talk about each weapon individually.

Mk 12

Officially the United States Navy Mk 12 Special Purpose Rifle, this gun has a sixteen-inch barrel, but is otherwise the same platform as an M 4. It fires a 5.56 x 45mm round from a thirty-round magazine. It can also be fitted with a twenty-round box.

U.S. Marines conduct live-fire training with an SR21 sniper rifle at Stone Bay in Marine Corps Base Camp Lejeune, North Carolina, March 15, 2011. The Marines are assigned to Battalion Landing Team, 2nd Battalion, 2nd Marine Regiment, 22nd Marine Expeditionary Unit.
(U.S. Marine Corps photo by Sgt. Joshua Cox)

Derived from what became known as the .223 cartridge and therefore smaller and lighter than most earlier military rounds, the 5.56 is not a preferred bullet to shoot someone with. It can take a few shots to put someone down, especially the drugged-up crazies we were dealing with in Iraq, unless you hit him in the head. Contrary to what you're probably thinking, not all sniper shots, certainly not mine, take the bad guys in the head. Usually I went for center mass— a nice, fat target somewhere in the middle of the body, giving me plenty of room to work with.

The gun was super-easy to handle and was virtually interchangeable with the M 4 which, though not a sniper weapon, is still a valuable combat tool. As a matter of fact, when I got back to my platoon, I took the lower receiver off my M 4 and put it on the upper receiver of my Mk 12. That gave me a collapsible stock and allowed me to go full-auto.

On patrol, I like to use a shorter stock. It's quicker to get up to my shoulder and get a bead on somebody. It's also better for working inside and in tight quarters.

Another note on my personal configuration: I never used full-auto on the rifle. The only time you really want full-auto is to keep someone's head down; spewing bullets doesn't make for an accurate course of fire. But since there might be a circumstance where it would come in handy, I always wanted to have that option in case I needed it.

Mk 11

Officially called the Mk 11 Mod X Special Purpose Rifle and also known as the SR25, this is an extremely versatile weapon. I particularly like the idea of the Mk 11 because I could patrol with it (in place of an M 4) and still use it as a sniper rifle. It didn't have a collapsible stock, but that was its only drawback. I would tie the suppressor onto my kit, leaving it off during the start of a patrol. If I needed to take a sniper shot, I would put it on. But if I was on the street or moving on foot, I could shoot back right away. It was semiautomatic, so I could get a lot of bullets on a target. It fired 7.62 x 51mm bullets from a

A Polish soldier fires a sniper rifle during a live-fire exercise at Forward Operating Base Warrior, Afghanistan, March 21, 2011.
(U.S. Army photo by Spc. David Zlotin)

twenty-round box. Those had more stopping power than the smaller 5.56mm rounds. I could shoot a guy once and put him down. Our rounds were match-grade ammo bought from Black Hills, which makes probably the best sniper ammo around.

The Mk 11 had a bad reputation in the field because it would often jam. We wouldn't have jams that much in training, but overseas was a different story. We eventually figured out that something to do with the dust cover on the rifle was causing a double feed; we solved a lot of the problem by leaving the dust cover down. There were other issues with the weapon, however, and personally it was never one of my favorites.

.300 Win Mag

The .300 is in another class entirely. As I'm sure many readers know, .300 Win Mag (pronounced "three hundred win mag") refers to the bullet the rifle fires, the .300 Winchester Magnum round (7.62 x 67mm). It's an excellent all-around cartridge, whose performance allows for superb accuracy as well as stopping power. Other services fire the round from different (or slightly different) guns; arguably,

U.S. Army Maj. Tracy Kreuser fires an XM2010 sniper rifle on Bagram Airfield, Afghanistan, April 8, 2011. The Army distributed the newly purchased weapon to sniper teams operating in Afghanistan. Kreuser was with the Intelligence and Sustainment Company, Headquarters and Headquarters Battalion, 101st Airborne Division. (U.S. Army photo by Sgt. Grant Matthes)

the most famous is the Army's M 24 Sniper Weapon System, which is based on the Remington 700 rifle. In our case, we started out with MacMillan stocks, customized the barrels, and used the 700 action. These were nice rifles.

In my third platoon—the one that went to Ramadi—we got all new .300s. These used Accuracy International stocks, with a brand-new barrel and action. The AI version had a shorter barrel and a folding stock. They were bad-ass.

The .300 is a little heavier gun by design. It shoots like a laser. Anything from a thousand yards and out, you're just plain nailing it. On closer targets, you don't have to worry about too much correction for your come-ups. You can dial in your 500-yard dope and still hit a target from 100–700 hundred yards without worrying too much about making minute adjustments. I used a .300 Win Mag for most of my kills.

.50 Caliber

The fifty is huge, extremely heavy, and I just don't like it. I never used one in Iraq. There's a certain amount of hype and even romance for these weapons, which shoot a 12.7 x 99mm round. There are a few different specific rifles and variations in service with the U.S. military and other armies around the world. You've probably heard of the M 82 or the M 107, developed by Barrett Firearms Manufacturing. They have enormous ranges and in the right application are certainly good weapons. I just didn't like them all that much. (The one .50 I do like is the Accuracy International model, which has a more compact, collapsible stock and a little more accuracy; it wasn't available to us at the time.)

Everyone says that the .50 is a perfect anti-vehicle gun. But the truth is that if you shoot the .50 through a vehicle's engine block, you're not actually going to stop the vehicle. Not right away. The fluids will leak out and eventually it will stop moving. But it's not instant by any means. A .338 and even a .300 will do the same thing. No, the

U.S. Marine Corps Lance Cpl. Daniel R. Ferree, right, a sniper with Weapons Company, 81 mm Mortar Platoon, 1st Battalion, 5th Marine Regiment, exchanges greetings with a group of Afghan children at the edge of a wheat field in the village of Sareagar, Helmand province, Afghanistan, April 16, 2011.
(U.S. Marine Corps photo by Cpl. Benjamin Crilly)

best way to stop a vehicle is to shoot the driver, and that you can do with a number of weapons.

.338 Lapua

We didn't have .338s in training; we started getting them later on during the war. Again, the name refers to the bullet; there are a number of different manufacturers, including MacMillan and Accuracy International. The bullet shoots farther and flatter than a .50 caliber, weighs less, costs less, and will do just as much damage. They are awesome weapons.

I used a .338 on my last deployment. I would have used it more if I'd had it. The only drawback for me was my model's lack of a suppressor. When you're shooting inside a building, the concussion is strong enough that it's a pain, literally. My ears would hurt after a few shots.

U.S. Marine Corps Gunnery Sgt. Tony Palzkill uses the mounted scope on his M110 semi-automatic sniper system to provide overwatch for his fellow Marines during a foot patrol at Sangin, Afghanistan, May 19, 2011. Palzkill was assigned to Scout Sniper Platoon, 1st Battalion, 5th Marine Regiment, Regimental Combat Team 8.
(U.S. Marine Corps photo by Cpl. Nathan McCord)

Since I'm talking about guns, I'll mention that my current favorites are the weapons systems made by GA Precision, a very small company started in 1999 by George Gardner. He and his staff pay close attention to every detail, and his weapons are just awesome. I didn't get a chance to try one until I got out of the service, but now they're what I use.

Scopes are an important part of the weapon system. Overseas, I used a 32-power scope. Additionally, depending on the circumstances, I had an infrared and visible red laser, as well as night vision for the scope.

As a SEAL, I used Nightforce scopes. They have very clear glass and they're extremely durable under terrible conditions. They always held their zero for me. On deployments, I used a Leica range finder to determine how far I was from a target.

Most of the stocks on my guns used adjustable cheek-pieces. Sometimes called a comb (technically, the comb is the top piece of the stock, but the terms are sometimes interchanged), the extension let me keep my eye in position when sighting through the scope. On older weapons, we would adapt a piece of hard-packed foam and raise the stock to the right height. (As scope rings have gotten larger and more varied in size, the ability to change the stock height has become more important.)

I used a two-pound trigger on my rifles. That's a fairly light pull. I want the trigger to surprise me every time; I don't want to jerk the gun as I fire. I want no resistance: Get set, get ready, put my finger and gently start squeezing, and it goes off.

As a hunter, I knew how to shoot, how to make the bullet go from point A to point B. Sniper school taught me the science behind it all. One of the more interesting facts is that the barrel of a rifle cannot touch any part of the stock: they need to be free-floating to increase accuracy. When you shoot a round, a vibration comes through the barrel, known as barrel whip. Anything touching the barrel will affect that vibration and, in turn, affect the accuracy. Then there are things like the Coriolis effect, which has to do with the rotation of the earth and the effect it has on a rifle bullet. This comes into play only at extremely long distances.

You live all of this technical data in sniper school. You learn about how far to lead someone when they're moving—if they're walking, if they're running, depending on the distance. You keep doing it until the understanding is embedded not just in your brain but in your arms and hands and fingers.

In most shooting situations, I adjust for elevation but not for windage. Simply put, adjusting for elevation means adjusting my aim to compensate for the drop of my bullet over the distance it travels; windage means compensating for the effect of the wind. The wind is constantly changing. So about the time I adjust for wind, the wind changes. Elevation is a different story. If you're in a combat situation, a lot of times you don't have the luxury of making a fine adjustment. You have to shoot or be shot.

Super Sniper in Iraq

By Craig Roberts

Marine Scout Snipers from the 3rd Battalion, 7th Marines shoot their way though enemy territory during the U.S. invasion of Iraq.

From an interview with USMC Sgt. Joshua Hamblin, Scout Sniper, 3rd Battalion, 7th Marines, Baghdad, Iraq, 2003

U.S. Marine Corps scout snipers with Regimental Combat Team 6 look out over Fallujah, Iraq, 18 September, 2007, from a bunker on Observation Post Sina'a.
(U.S. Marine Corps photo by Lance Cpl. David Castillo)

A Long-Awaited Call

I was in my office conducting an interview when the call came in. When I answered, I could barely make out the caller due to static and breaking up of his sentences. Finally, after repeated questions that reminded me of having to try and call in fire missions in Vietnam on an antique PRC-10, I alerted to the fact that the caller was Sergeant French, using a SAT phone, calling from Baghdad!

I had been working on my latest sniper book, *Crosshairs on the Kill Zone*, and had been jumping through journalistic hoops with the Marine Corps in an uphill battle to find and interview Marine scout snipers in Iraq. Due to the fact that the initial invasion had happened just weeks earlier, security was extremely tight and I had been told by public affairs officers from Camp LeJeune to Quantico to Headquarters Marine Corps that no one could talk to me or anyone else. But I kept climbing the PR ladder until I finally struck pay dirt. The Commandant's office authorized the interview and the call finally came.

Sgt. French advised that they would call the next day on the Sergeant Major's personal SAT phone, and I would be speaking with one of his top Scout Snipers in 3/7, Sergeant Joshua Hamblin. That interview revealed the following incredible, danger-packed, harrowing two-week sniper adventure.

Carnage Created from Atop a Roof

"Our job was to provide reconnaissance and surveillance, and, when necessary, precision fire in support of the operations. We knew our job, but we never realized just how effective we could be, and how much carnage a pair of snipers could create until the day we crawled onto a rooftop at the Al Rashid Military Complex near Baghdad.

"We, the 3rd Battalion, 7th Marines, had entered Iraq from Kuwait, traveling north in HMMWVs as fast as the terrain allowed. In my vehicle, I carried an M 16 for suppressive fire and my M40A3 sniper rifle for precision fire. I kept my sniper rifle right next to me in case it was needed. Whenever we came up to a road block, I was able to engage the Iraqis manning the roadblock with, as they say, "precision fire." That means lining them up and knocking them down.

U.S. Marine Corps Sgt. Adam Storey, a Scout Sniper assigned to Team 4, Company G, 2nd Battalion, 24th Marine Regiment, Regimental Combat Team 1, uses the scope atop his M40A1 sniper rifle to observe a cement factory near Bahkit, Iraq, July 3, 2008.
(U.S. Marine Corps photo by Cpl. William J. Faffler)

We Never Missed Our Target

"As we moved through Iraq, we were attached, or "punched out" to various companies in the battalion for stalking out and doing observation and reconnaissance missions along the way. Sometimes the companies would have to stop for a few days and, while doing so, the Scout Snipers' mission was to move out ahead and watch the population for suspicious activity. In one area, we saw no Iraqi soldiers, but plenty of young males of military age with short haircuts. It was obvious that these guys had discarded their uniforms and were trying to blend into the population, but there wasn't much we could do about it right then. The 'rules' called for us to ignore those who weren't armed or did not attack us. But if we saw an Iraqi who was armed, we were to take him down. So, any Iraqi with an AK was a target, and we never missed.

"Eventually, after fighting our way north through villages, towns, roadblocks and ambushes, we began drawing missions to support special operations raids that were looking for Baath party members and former members of Saddam's regime. We did this by covering the raiding party with our sniper teams to protect them as they did building entries.

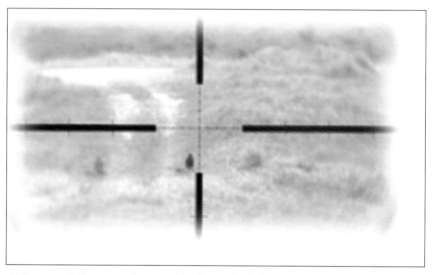

A view through the scope of an M40A3 sniper rifle. Marine Sgt. Joshua Hamblin saw numerous Fedayeen gunmen this way before he whacked them.
(Photo by Lance Cpl. Tyler Hlavac, USMC)

Encumbered by Cumbersome Gear

"All of this time we were fighting the elements, and it was miserable. Dressed head to foot in heavy chemical suits, we could barely breathe while carrying our gear, wearing Kevlar vests and helmets, plus weapons and ammo. We had to adapt as we went because no one had fought for an extended period like this, with all this gear, for this long.

"But worst of all was the lack of rest and the filthy conditions. There was no water for bathing, so whenever we found water or a broken pipe, we took time to wash down. As for sleep, we usually were lucky to get two hours of sleep before we were punched out to work another mission. It was not unusual to go two days or more without sleep. In these cases, we ran on pure physical discipline and will power.

"We basically worked two types of operations—one in the bush and another on buildings. In the bush operations, we set up hides, using camouflage and concealment, and observed the population for targets or threats. In the built-up areas, we hopped up on rooftops to provide covering fire or area denial missions, where we denied the area to the enemy by covering it with direct fire.

U.S. Marine Corps 1st Lt. Bryan J. Abell (left), Scout Sniper Platoon leader for 3rd Battalion, 1st Marine Regiment, and Cpl. Jason P. Abell, a machine gunner with 3rd Battalion, 25 Marines, recently met at Camp Hit, Iraq. At the time the photo was taken, Jason was on his way home after serving in Operation Iraqi Freedom while Jason was starting his second tour in Iraq. (U.S. Marine Corps photo by Cpl. Adam C. Schnell)

We Reach the Al Qaeda Camp

"We finally made our way into the facility of Salmon Pak, where we had expected to meet a lot of Iraqi resistance. This is the location of the al Qaeda terrorist training camp, with the fuselage of an airliner for training hijackers. But instead of fierce resistance, we found destroyed buildings and burned out hulks where the air cover had done its job. There were scores of empty vehicles, empty fighting holes, and more young men with short haircuts, but no fighting.

"We continued north and, unlike some of the areas that were open desert, this was the Tigris River valley, which was lush with vegetation. This provided us with much more suitable camouflage and a much better environment to conduct operations in. But it was really now starting to get hot. We found ourselves taking more breaks just to air ourselves out. You would take off your chemical suit and just be drenched underneath. It was as miserable as having your own personal sauna bath that you wore—and couldn't get rid of.

"A week after leaving Salmon Pak, while striking toward Baghdad and living on only two hours of sleep a day, the word came down that we could discard our chemical suits! This was the most welcome news we had had in a long time. Finally we could work in our desert utilities and fight like normal marines. It felt like the weather had cooled by several degrees and was a major turning point in the war for us.

Positioning at Al Rashid Military Complex

"As we neared Baghdad, we ran into more minefields, and we could hear a lot of shooting going on, plus a lot of artillery. Then, as we approached the Al Rashid Military Complex, we could see breaches in the wire and more mines scattered around, but still did not run into any significant resistance. It was as if no one was there, but they might come back if we hung around for a while.

"Taking over the complex, we set up positions to defend the place against a counterattack. Then, my partner, Sergeant Owen Mulder, and I were assigned an area of responsibility to set up a Scout Sniper position on a one-story house on the northeast side of the complex. This was what is known as an overwatch position, which allowed us to observe the approaches to that area from that direction. The farthest engagement range appeared to be about 750 meters.

"Though there were two companies at the complex, one on each flank about 150 meters away, the area we covered was totally up to just my partner and me. Between us, Mulder held his M40A1 sniper rifle with the standard issue Unertl 10X scope, and I had an M40A3 sniper rifle with the ANPVS-10 day and night sight. The ANPVS-10 is a third-generation night image intensifier with daylight capabilities. We could work the position night and day, which is exactly what we would find ourselves doing.

"We got up on the roof at night and I turned on my night scope. Mulder had night vision goggles, so we both could see the streets to our front in a clear ghostly green panorama. For American forces, our night vision capability was second to none, and it gave us a giant advantage over troops not similarly equipped.

A night vision shot of U.S. Marine Scout Snipers, one with an M40 bolt-action sniper rifle, assigned to Team 4, Co. G, 2nd Battalion, 24th Marine Regiment, Regimental Combat Team 1, observing a cement factory near Bahkit, Iraq, July 3, 2008.
(U.S. Marine Corps photo by Cpl. William J. Faffler)

"I Got Targets"

"After two hours, we spotted a pickup truck approach, stopping about 400 meters short of coming into the military complex. As we watched, we could see that the truck had a bunch of people in the back. Then, watching closely, we saw them begin jumping out of the truck, each carrying an AK. My pulse began to race.

"'I got targets,' I said to Mulder. 'See 'em?'

"'Yeah, let's go for it.'

"Immediately I decided to take a shot, and as I did I could see the driver jump out and go around the back of the vehicle. I lined him up in my sight and squeezed the trigger. He went down like a sack of potatoes. Mulder sighted in and fired, too, and another Iraqi went down.

"We just began lining up targets as fast as we could and taking the shots. As we did, other soldiers tried to drag the bodies off, but when they did, we took them out, too. It was like a shooting gallery. The more shots we took, the more targets seemed to appear. The adrenalin rush was indescribable. But training and discipline kept us pacing ourselves throughout the shots, taking over where runaway excitement wanted to rule. Spot a target, sight in, squeeze, watch the shot, work the bolt, pick another target. Over and over.

All Was Quiet Again in the Kill Zone

"Then the next thing I saw was the Iraqis dragging bodies into the pickup truck, driving around the corner out of sight. All was quiet again in the kill zone.

"I looked at my watch. About an hour before sunrise.

"Nothing happened until after the sun came up. The Iraqis knew by then that trying to move at night would do them no good. In fact, to even up the odds a bit, they would have to move in daylight so they could see as well as we could. This would be another major mistake on their part. They simply did not understand the capability of a well-trained and equipped sniper team.

"We observed the street for about an hour. Then I spotted two individuals enter our field of view, trying to cross the street. One was carrying an AK. Good enough.

"'You got these guys?' I asked.

"'Got 'em. You taking out the guy with the AK?'

"'Affirmative.'

"I estimated the range: 450 yards. I led him slightly and fired. He went down, but he wasn't killed immediately. His partner took off and didn't come back. The first guy stopped moving about thirty seconds later.

"Our rules of engagement called for us not to shoot anyone who was not armed or did not pose a threat. But if they carried weapons, they were bought and paid for. Also, if we had a suspicious vehicle that came and went several times, we could engage it.

"About ten minutes after that, a truck pulled out of one of the side streets. Then fifteen minutes later the same truck came back with the same guy driving, followed by a second truck. Each was carrying seven or eight guys in the back.

"'Watch these guys. They're probably armed.'

A Marine sniper and two others stand in front of an Abrams tank.

They Jumped Out, Armed With AKs

"Then, as we watched through the scopes, both trucks stopped and the riders all jumped out. Each had an AK.

"My partner and I didn't even have to say anything to each other. I took the closest truck and he took the far truck. We started shooting people. As fast as we could work the bolts and acquire targets, we squeezed the triggers and an Iraqi went down.

"One was dropping right after another, just dropping like flies. We ended up shooting every one of them—all thirteen—right there, except one guy who managed to get into one of the trucks by shoving the dead driver out of the way and trying to back out of there. But we fired at the truck and I couldn't believe what happened next. For some reason, we hit something critical because the gas tank exploded in a ball of flame, just like in the movies! I thought that cars and trucks blowing up when you shot at them was just Hollywood, but this one did exactly that.

"The truck was still moving, so we both shot through the driver's side of the cab. Two 175-grain boat-tail hollow points blasted through the thin sheet metal of the cab and killed the driver as it coasted around the corner. He would cook in silence, just out of sight, for many minutes.

"So now we have sixteen bodies littering the street, plus a burning wreck and an abandoned truck, all due to two snipers. It looked like a major battle had been fought there, but it was just Mulder and me, and we hadn't even fired a full box of ammo yet. The amazing scene of a vehicle blowing up would happen two more times over the course of that day.

"Later in the morning, as I watched the street through my scope, I saw an Iraqi soldier turn the corner and walked toward us. He was in full uniform, wearing a helmet and vest and carrying a sidearm. It was as if he was just showing up for work that day and didn't know we were there. He walked toward us along a row of light posts. As he passed each post, I set the scope on him and followed and watched as he lessened the range.

"About this time, an old man came out from one of the houses alongside the road and began gesturing at him, trying to tell him to get out of there. But the soldier didn't seem to understand what he was trying to say and kept coming.

"Mulder and I kept shifting our attention from the soldier to the old man, and back again. It was like watching a comical movie, like 'hey, get out of here. They'll kill you!' and the soldier saying 'Who? Who'll kill me? What are you talking about, old man?'

Civilians Watch a Horror Show

"Finally, as I watched the soldier shrug his shoulders one last time, I squeezed the trigger and could see the vapor trail of the bullet. It flew straight and impacted him in the center of the chest, like watching a movie in slow motion. I could see the dust fly off his shirt, then saw the expression on his face of pure shock. He was completely surprised. He looked down, then slumped over and fell to the ground. It was my best shot and I could see everything. The bullet path, the impact, the expression on his face. It was perfect.

"He tried to crawl out of the kill zone, but it was too late for him. He only moved for a few feet, then became very still.

"The local civilians were very interested in what was happening and watching all the action became an amusement to them. After each episode of firing and dropping soldiers, the civilians would come out, look around and laugh and smile, then look over toward the base and wave at us. It seemed almost like they were spectators at a sporting event.

"By this time, there's seventeen bodies laying out there, with two shot-up pickup trucks, one still smoking and smoldering. It was like a scene out of a *Rambo* movie, but it was real. And here were all these people just lined up on the side of the street looking at the carnage like it was a day at the market.

"Suicide by Marine"

"Just then a white pickup truck pulled out of a side street, and in the back was a typical terrorist with a black ski mask and black clothes, holding an AK. The truck turned and sped toward us. He was obviously a Fedayeen and was hell bent on 'suicide by Marine.' As the truck careened down the road, with this guy holding on for dear life, I raised my head to look at him at the same time he looked up at me. I went back down on the scope and worked to keep on target as the pickup truck bounced along the road. I found a lead angle, aimed high and took the shot. It was a dead slam hit and he went down in the back of the truck. We didn't get a chance at a second shot at the driver because he turned away and raced around a corner out of sight. But the guy in the black pajamas was down and finished.

"Eighteen.

"We continued to watch vehicles move in and out of the area, always trying to memorize them to see if they were repetitious. It seemed there were certain color vehicles that were used by the military and we started learning what to watch for. Most were Toyota light-colored pickup trucks. We noticed that the trucks were making runs into a nearby neighborhood carrying stuff in the back of the trucks going in, then coming out empty. It became apparent that they were doing a supply run, stocking a strongpoint with weapons and ammo. It was time to put a top to this.

"The rules of engagement were that you could shoot a man with a weapon, or a vehicle carrying weapons by shooting the tires out or otherwise disabling the vehicle. One red Toyota car had made one too many trips and was now on the hit list.

A Ball of Flame

"The fourth time we saw the car, Mulder and I opened up. Mulder hit the driver and I hit the back of the trunk. Then the whole damn thing exploded! The whole trunk blew up in a ball of flame, then the car coasted around the corner. A few minutes later, all hell broke loose as a trunkload of RPG rockets cooked off. We looked at each other as we listened to the distinctive *whoosh . . . bag . . . whoosh . . . bang* sound of the rockets igniting, taking off in odd directions, then exploding as they made contact with buildings, streets, and anything hard.

"It seemed like two 175-grain bullets had started a small war out there. By this time it dawned on us that we had not received any return fire during the engagements. We were in total control of the position, and there was nothing the Iraqis could do about it. If they tried to show themselves to engage us, we killed them. They had no chance to set up, locate us, then put fire on us in the time it took for us to see and shoot them. Plus we had cover and a much better long-range capability. We were trained Marine snipers and they were overzealous fanatics with little marksmanship training. There was no comparison. It was like the old joke of 'don't attack that hill, it's a trick. There's two marines up there.'

"Any time we saw an individual with a weapon to our front, he was history. But we did receive incoming fire from the flanks and couldn't do anything about it except hope we didn't get hit by "collateral damage." The line companies couldn't control all the individuals they were having contact with and several firefights were erupting on both flanks as Marine units ran into pockets of well-armed Iraqis. Rounds flew everywhere from those actions, and many flew right over us. Though we had total control of the area to our front, we had no control of what dangers lay to each side.

Hurt A Marine—Die!!!

"While we were there, one of our snipers from the Scout Sniper Platoon, Sergeant Aaron Wintterle, who carried a Barrett .50 Special Applications Scoped Rifle, was manning a vehicle check point with one of the line companies when an Iraqi suicide vehicle approached. As Wintterle trained his .50 SASR on the vehicle, it exploded. A piece of shrapnel hit him in the face, breaking his jaw and putting him out of action. Immediately, his partner, L/Cpl. Jacob Heal, who was a new guy under training, jumped on the weapon and spotted one of the Iraqis who was involved trying to run back down the road. The marine took aim and lit him up, blowing out his chest.

"Meanwhile back at the Al Rashid Military Complex, rounds came past us from the sides, along with RPG fire and explosions as close as fifty meters away. This became the routine throughout the day as sporadic fighting continued on the flanks.

"But our job was to keep our front secure. Every time we found an armed man, we took him out. When we saw armed trucks, we took them under fire, as well. It was really weird to spot a pickup with a .50-caliber mounted in the back, then shoot the driver and gunner, then watch the truck just keep going like it's driven by a ghost.

"When we saw weapons inside a car or truck, we engaged both the passenger and the driver. Our method was simply 'whoever spotted the weapon first called the shot.'

"'I got RPG rounds. I'll take the driver. You take the passenger.'

"The other guy then would take out the passenger at the same time. This was the great thing about having two sniper rifles there, since we could fire on two targets at the same time.

"Our body count continued to climb throughout the day.

Adrenalin-Packed Super Snipers

"We were up six days before we got away from the roof and got four hours sleep. This was counting the three days movement through Salmon Pak, then the three days on the roof. Then it was back to the same position. Then we were up for another seventy-two hours constantly, moving or staying on the scope. Watching and shooting. We were running on pure adrenalin.

"It got to the point where if an Iraqi entered our area, they took off their shirt, pulled off their white T-shirt and waved it like a truce flag until they got out of our area, then put their shirt back on. The word was out, and no one wanted to come into our kill zone to die for Saddam anymore.

"Still, we continued to watch and wait. We took turns, an hour on the glass, an hour off the glass. Though it was fatiguing, we stayed alert. The adrenalin and the anticipation that anything could happen any second kept us on the keen edge of awareness. In fact, I was so pumped up on adrenalin that I could have stayed up for a week.

"By the time we left this position, I had seventeen kills and Mulder had fifteen. Thirty-two total kills in one spot!

Olympic Stadium—Final Destination

"With the main part of the Iraqi resistance now crushed, we entered Baghdad proper and pushed in toward the city center. As we moved down the highways and streets, we could see vehicles on side streets that had been taken out the day before and were smoldering and had rounds still cooking off. We ran into sporadic firing, both small arms and RPGs, but we aggressively moved through anything we came up against and pressed on. Our final destination was the Olympic Stadium.

"The Republican Guards had told everyone that the marines would kill everyone, and that we would eat them or some such nonsense. But when we moved through neighborhoods and people saw that we treated them well, and that we had finally run the Republican Guard off, they came out in the hundreds of thousands and began cheering us, dancing in the streets.

"Still, even after we achieved positions at the Olympic Stadium and began to consolidate, we received sporadic incoming rounds. It was becoming obvious that it would take a long time to totally pacify Baghdad.

"For two weeks, we continued to do our surveillance and target acquisition missions, then pulled out to set up a more permanent base camp. By this time, most of the Iraqi Army had thrown their weapons down, and basically said, 'Okay, you win.'

"Mulder and I would not see a more target-rich environment than what we had at the al Rashid gate."

A Marine Mobile Sniper Strike Team

Marines Shoot Way Through the "Baghdad Two-Mile"

By GSgt. Jack Coughlin, USMC (Ret.)

A member of Scout Sniper Platoon, Weapons Company, Battalion Landing Team, 3rd Battalion, 2nd Marine Regiment, 22nd Marine Expeditionary Unit, sights in on a target with an M40 sniper rifle during a live-fire range exercise at Fort Pickett, a Virginia Army National Guard Maneuver Training Center. The Marines of BLT 3/2 conducted off-site training, the first major exercise of a six-month pre-deployment training cycle for the 22nd MEU.
(Photo by Cpl. Theodre Ritchite, 22nd Marine Expeditionary Unit.)

Gunnery Sgt. Jack Coughlin joined the Marine Corps in 1985, after an injury derailed his dreams of becoming a major-league pitcher. He would go on to become one of the Marines' elite Scout Snipers, and his service included a tour of duty in Somalia. However, he would make his mark during the liberation of Iraq, where he had the chance to pioneer the Mobile Sniper Strike Team concept. Here, he describes one of the many engagements that Marines had with Saddam's thugs while on the road to Baghdad.

—Ed.

A Town Called "Hell"

Death was just beyond my windshield. Bullets whanged against metal and shoulder-fired anti-tank (RPG) rockets zipped toward us. Marines were running and gunning amid the chattering pops of small-arms fire, and machine guns were firing full out, with a heavy, rhythmic stomp. Artillery shells exploded and shook the ground.

On the map, this town was called Az Zafaraniyah. But on that April morning, it was hell for the U.S. Marines, a raging, brutal fire-fight that gave this new generation of jarheads a taste of what the Corps had faced on Guadalcanal in World War II, at the Chosin Reservoir in Korea, and at Khe Sanh in Vietnam. It was not pretty enough to appear on TV, for our job was to heap casualties on the defenders, fast and with unrelenting force, and force them to with-draw. So time slowed down and I went on a killing spree.

McCoy had been thoroughly ripped that I was not out on the front edge of the attack, dominating the rooftops when the fight started, and he had barked out on the open radio frequency, "Listen, I want Coughlin and his rifle up here right now!"

I grabbed the mic and confirmed that I was on the way. "Good," the colonel grumbled. "See you soon." We'd explain after the shoot-ing stopped, but right now there was work to be done.

Game On!

"Game on, boys!" I yelled. "We're back in it."

Snipers with Weapons Company, Battalion Landing Team, 2nd Battalion, 5th Marine Regiment, sight in their M 40A3 sniper rifles at the Camp Hansen Range.
(Photo by III Marine Expeditionary Force Public Affairs)

Lance Cpl. Marco S. Buehler, a Scout Sniper assigned to Battalion Landing Team 2/6 of the 26th Marine Expeditionary Unit, fires his MK 11 sniper rifle from a CH-46 Sea Knight helicopter while flying over the Gulf of Aden. The snipers train by firing at floating targets from different places inside the helicopter. The 26th MEU is part of the Iwo Jima Expeditionary Strike Group and is deployed in the U.S. 5th Fleet area of responsibility.
(Photo by Cpl. Aaron Rock, Navy Visual News Service)

Our trucks were rolling almost before I finished telling the boys to get cracking, and we went from loafing to combat-ready in a heartbeat. As the two Humvees broke out of line and gathered speed, Officer Bob appeared in the road, waving his arms. Casey told his driver, "You work for me, not him. Do not stop." His Humvee dodged the captain and kept going.

I, however, wanted a few words with the lad, so I had the Panda pull to a quick stop. The officer rested his hands on the edge of my window and asked, "Where you guys going?"

I pointed north, toward the gunfire. "Right up there. You told me the colonel didn't need me in this attack. You told me there was not even going to be an attack today." My voice was shaking with anger as he dumbly nodded his head.

"Aren't you monitoring Tac-1?" I asked. "Colonel McCoy just personally ordered me up to the front, and he's more than a little pissed that I'm not there already. I'll let you explain it to him later. . . . Sir."

The Panda gunned the engine, almost dropping Bob into the road as he stammered something I didn't bother trying to hear.

Orders to Shoot any Threat

As our Humvees careened through the narrow streets, I scrubbed my mind clear and loosened my muscles in preparation for combat. My fingers walked along the length of my sniper rifle, unconsciously checking it for imperfections with a sense of touch as accurate as a concert pianist playing Mozart on a familiar Steinway. The gun's magazine carried four rounds, and I slid the bolt back slightly and stuck my finger inside the raceway to check for brass.

There was already a round in the chamber. I sighed, content and ready.

Smoke floated into the sky above the flat rooftops of Az Zafarani-yah. Casey worked the radio as our Humvees sailed down a street of low-walled homes already pockmarked by bullet strikes, and he let India Company know we were coming up behind them. We were in a precarious position, because we were not directly affiliated with any of the platoons or companies, and we could easily become targets if we showed up unexpectedly on their dirt. "If anybody poses a threat to you, you kill them," McCoy had ordered his combat teams right before the attack. Violent supremacy would rule this day, and there was a chance that we might get zapped by our pals.

The Baghdad Two-Mile Marathon

We stopped at a gate, at which a couple of Amtracs were firing like crazy, jumped from the trucks, and grabbed our gear. I would carry only my rifle, the sniper log book, a pistol, and four quarts of water. Casey toted a low-power handheld radio, maps, water, and his own weapons. Then we had to break up our team. We picked Tracy and Newbern to come with us. The other guys would provide support with the guns on the Humvees. Everyone picked up a twenty-round box of the precision-made ammo to feed my rifle. At a hundred yards, those special bullets would hold within a one-inch circle; at a thousand yards, within a circle ten inches in diameter. The average human head is about twelve inches in diameter.

The guys in the Amtracs could tell us little about what was going on, other than that there was a lot of shooting, so we cautiously stepped around them and entered the city. Newbern took point. I came next, then Tracy, with Casey hustling along to provide rear security and radio communications. I could hear him telling India Company that we were heading for the rooftops. It was the start of an unofficial marathon that came to be known as the "Baghdad Two-Mile," an event that will never make it to the Olympics.

As soon as we were in the clear, we broke into a gallop for the built-up north side of the road leading to the bridge and rushed inside a two-story building that the India Marines had just cleared. We pounded through the darkness and up the stairs into the hot sun-shine that blazed down on the roof. No wind at all. Great shooting weather.

A two-foot wall encircled the top of the small building, and I pulled a cinder block over to a corner and sat on it, arranged some gear on the top of the barrier, and pushed the rifle into a sturdy position. Then I locked into a tight shooting posture and glassed the area, looking slowly north of our position and blinking away the sweat that stung my eyes. The MOPP suit was flaming hot.

A Matter of Professional Honor

My scope was drawn to a small blue-tiled minaret that rose above the surrounding brown buildings, where I saw a flicker of movement. Somebody was hiding behind a wall high above the street, an enemy fighter in civilian clothes with an AK-47, and I saw him peer down into the maze of streets below. The bastard was doing the same thing I was doing—looking for targets. I had found an enemy sniper, so this instantly became a matter of professional honor: I'm better than you, rat.

The last brief for Coughlin's unit prior to crossing the line of departure is given by Lt. Col. Bryan P. McCoy, the battalion commander.

Moving fast, to get him before he could open fire on the Marines, I painted a quick laser on his position, dialed in exactly 343 yards, and planted my crosshairs right on his chest. All the while, my mind was unconsciously wheeling through the sniper's mantra of S—Slow, Smooth, Straight, Steady, Squeeze—and the rifle seemed to fire on its own. My bullet bored perfectly into his chest, and its heavy mass penetrated his major blood-carrying organs, crushing and destroying tissue. That created a hole that is called the "permanent cavity," and then the bullet expanded, sending small, jagged fragments spinning off in erratic paths that shattered his organs. Had to hurt. I watched him slump into a fetal position. Although his body might twitch for another few seconds, this guy was dead.

I'm Untouchable

I was in my zone. The protective presence of Casey, Newbern, and Tracy around me, and the advancing marines and armor down in the streets, allowed me to concentrate totally on being a pure shooter, an ethereal feeling of being untouchable and able to reach out and control the destinies of other men.

McCoy was off somewhere ramrodding the entire battlefield, with enough radios to talk to anybody on the planet. The company and platoon commanders were making sure where the rifles were pointing, platoon sergeants were hollering orders, and fire team leaders were pushing marines into exact positions and kicking butts to make them run faster. Casey had binos at his eyes and a radio glued to his ear to guide our little team. Tracy and Newbern were in nearby protective positions. These were all well-trained military personnel who understood the grand plan of battle unfolding that morning as we bashed into the river town. The great dance was in full swing.

I didn't have to worry about that shit. I was merely a destroyer of men.

I pulled the bolt back and reloaded, oblivious to what was going on around me. That was someone else's job, and if I needed to know something, they would tell me. As Casey later explained, "Unless I absolutely had to get into his zone, I left him alone. You don't want to mess with a man's zone, especially when he's killing people and doing good things."

My eyes seemed to magnify things even without using the scope. New smells drifted to me over the rooftops, and my hearing gathered all kinds of sounds while my brain filtered out the noise and turned down the volume, distinguishing one type of explosion from another or the whine of a passing bullet. Things seemed to slow down, and the adrenaline helped me move and think five times faster than normal. It is an inexplicable feeling that comes to warriors in the heat of a fight, and it has been described since the dawn of man. It is a cliché, it is mystical, and it makes no sense at all, but, by God, it is true.

Four minutes after I took out the sniper, Daniel Tracy called, "Boss, I got something out to the northwest." He verbally walked me across the rooftops, like a stranger in town giving directions to another stranger. *See that funny-looking building with the blue flower box? Up above that, the open window with the green curtain? Look left to the doorway.* I found the gunman Daniel had spotted, did a range check, squared up on the target, which was crouched half-seen 411 yards away, fired, and watched my bullet strike home and efficiently do its grim job at the far end of its parabolic flight. Another enemy soldier lay dead. I reloaded.

Seeing the Elephant

Casey was a different man. All of the apprehension and curiosity that preceded his first firefight were gone, replaced by the calmer mien of someone who had smelled the smoke, heard the bullets, and knew what do to. In the warrior's world, we called dramatic change "seeing the elephant." Once you saw it, you never forgot it. He listened to the position reports over the radio as India Company's grunts continued clearing the west part of the city. "Let's go," he ordered, stuffing his maps into his pack. "We stay up here any longer, the fight will pass us by."

We hurried downstairs, hollering, "Friendly coming out!" to prevent some skittish Marine grunt from lighting us up. India's commander had been so obsessed with training his boys to fight in an urban environment that some people had thought him a bit of a nutcase. But now the training was paying off as they worked efficiently, house by house, through this dangerous warren of homes, shops, and buildings. It was nice to be within their security bubble, for death could be lurking in the sewers, on the roofs, or in the bushes, and the grunts were constantly yelling to keep track of each other in this urban abyss.

Out in the street, mortar shells were detonating ahead of us, and India's Amtracs moved up the road, flanked by marines on foot. At each window, they drew figure-eights with the muzzles of their M 16 rifles, shooting quickly at any suspicious movement within. Individual fire teams handled different levels of buildings. To go around a corner, two men would set up, with one kneeling in front and the second resting his hand on the shoulder of the first. When the hand squeezed the shoulder, both men would pop around the corner with weapons ready to shoot. India had practiced the techniques for hours on end, and their attack moved inexorably forward with a smooth fluidity despite the continued incoming fire.

A Carpet of Gunfire

We jogged down the road, moving from one squad to another, dashing across open intersections as bullets whizzed and smacked around us. The bad guys were laying a carpet of gunfire.

There are definite things to look for when choosing a building for a sniper overwatch position, but we didn't have time to run the checklist. We just needed to get up on a roof somewhere, fast, and set up a killing field. A hundred yards down the road, we found a likely spot and dashed upstairs, but by the time we got there, the battle had already moved past. We needed to jump farther and faster, so we pounded back downstairs and rejoined the fast-paced assault.

We were sweating profusely in our MOPP gear and panting with exertion by the time we found a good building 350 yards down the road and had an India rifle squad clear it for us. Up to the roof we went, where we picked up a passenger, Italian photographer Enrico Dagnino, one of the Jackals. They had unfettered access to this battle, and Enrico wanted to follow us. Fine, we told him, take your pictures, but just do what we tell you, and you might live long enough to see them published. He did not explain that he had bounced from war to war for years and had seen more combat than any of us.

Anti tank rockets were swooshing all over the place, exploding overhead and raining shrapnel; the rattle of enemy small arms and machine gun fire was increasing, and the return fire from the Marines was deafening. The boys took security positions in the corners of the roof, and Casey contacted the battalion headquarters to let them know we were moving across organizational boundaries.

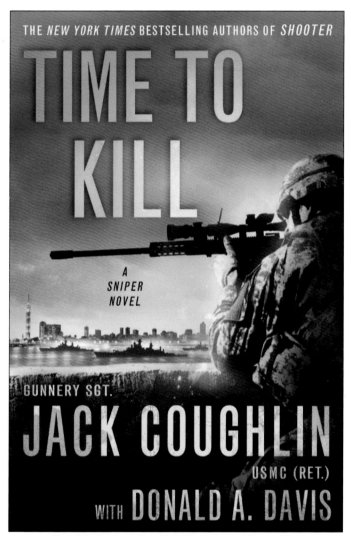

THE *NEW YORK TIMES* BESTSELLING AUTHORS OF *SHOOTER*

TIME TO KILL

A SNIPER NOVEL

GUNNERY SGT.

JACK COUGHLIN

USMC (RET.)

WITH DONALD A. DAVIS

Coughlin is also a New York Times *bestselling author.*

Like Paper Targets in a Carnival

Once again, I settled behind one of those low walls that ran the circumference of the roof, got the rifle in place, checked that I had a full load of ammunition, and took a look around. Three minutes later, I spotted a guy atop a two-story building who was firing an AK-47 and had an RPG strapped across his back. This dude definitely had to go. According to the laser, he was 550 yards away, and in this cluttered urban environment, somehow there remained a clear shooting lane between us, an open visual channel that yawned between the buildings from me to him. I hit him three inches below his throat and watched him sink onto the roof like a deflated balloon. The jackass had gone up there alone, with no security, and allowed his attention to be diverted elsewhere, away from the direct threat to himself.

I jacked in a new round without removing my eye from the scope, and quickly found another target in the open. The soldier was standing atop a chicken coop on the tin roof of a garage only 230 yards away, so close that his form almost filled my scope. He had his rifle in his shoulder and was popping away at marines, so I blew him away. It was as easy as shooting a paper target in a carnival midway. A bright red flash washed over his face when the round hit him in the mouth, and his head snapped back as if he had been tagged with a heavyweight boxer's left hook. The backward momentum snatched the rifle from his hands and knocked him not only off his perch but also clear off the tin roof. The body was limp when it hit the ground.

Mobile Sniper Strike Team

I reloaded, took a deep breath, and swept the area again, finding nobody else to shoot. I was still in my zone, emotion suppressed, brain engaged, my actions virtually robotic. My concept of a Mobile Sniper Strike Team—wheels to get to the area of action, then roaming at the front of the advancing forces, guarded by an experienced security team—was getting a thorough workout. My reach was hundreds of yards in front of the advancing troops, and I was sowing disarray and confusion. The idea worked!

Seven minutes after we bagged the guy on the garage, as we were getting ready to leave this building and move forward again to leapfrog the battle, an enemy soldier wearing the snazzy tan uniform and red beret of the Republican Guard walked into the middle of a street, almost as if he were on parade. From only 324 yards away, I spent thirty seconds examining him in detail and waiting to see if anyone else would join him in the open. He was calmly walking around as if he, instead of me, were the king of the world, and in his right hand he carried an AK-47 that looked almost new. Then he turned around, and that was his death notice, for it appeared that he might be leaving. I had the crosshairs precisely between his shoulder blades, and my bullet sent him slumping to the ground like Jell-O falling out of a mold.

Sniper One

By Sgt. Dan Mills

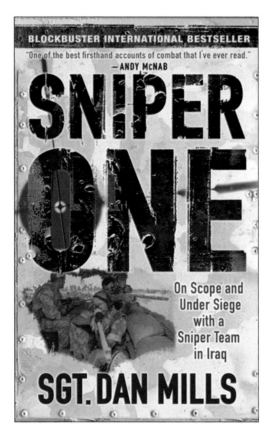

Desperate for Relief

All in all, by the end of the siege's second week we were in need—
desperate need—of any form of relief.

We got it.

Captain Curry had mentioned to me that there might be another
sniper pair coming our way from somewhere. When they turned up
one day in the middle of a big firefight, it was totally out of the blue.
Nobody quite worked out how they'd got there either; not even Captain Curry.

I was over in the Pink Palace when I got a call on the PRR [personal role radio] to come up to the Ops Room. Curry wanted to see me.

"Oh, hi Dan, thanks for coming over. I want you to meet these two chaps. They're a sniper pair from the . . ." he paused, as if considering the options, then continued, "er, the Royal Marines. They've come to help us out here for a little bit."

The first bloke extended his hand. He was in his late twenties, clean shaven, with mousy blond hair and a West Country accent. His haircut was the normal regulation short back and sides and he wore the usual British military combat fatigues.

A Cockney Named Just Buzz

"Hi, Marine John Withers."

Then the second stepped forward.

"Hello, mate, I'm Buzz." He was a cockney.

Interesting. First name only. And Buzz didn't wear any rank either.

"Dan Mills. Good to meet you."

Buzz looked nothing like us. He was older and shorter than John, in his thirties and stood at about five foot seven, as well as unshaven, with at least three days' stubble. He looked scruffy, with just a dark T-shirt on and a thin blue flak jacket over it, and a pair of civvy boots that weren't desert colored. In fact, the only military thing he had on was a very dirty pair of desert combat trousers.

It took me about five seconds to work out he wasn't Royal Marines. Might have been once, but not any more. He also carried a bloody large valise over his shoulder that was almost as tall as him and looked extremely heavy. What was in that?

Despite Buzz's appearance, both men were very polite and professional. "Can you give us a bit of a show around? We'd be grateful." said Buzz.

"Pleasure." I was happy to have any help we could get.

A Rooftop Visual Tour

Chris and I showed them all the positions we were using, and gave them a visual tour of the city from the roof. Buzz asked if it was okay if they worked from Rooftop sangar [firing position].

"Be my guest, mate; shoot from wherever you want."

"Thanks."

"Just one thing. John is, but you're not Royal Marines, are you."

Buzz just smiled. I'd come across a few of his type in my time. Always the same. They don't say, and you don't push them. You don't need to. Everybody knows the game.

Buzz had come down from Baghdad. He was a sergeant with his unit. He was too polite to say it, but it was obvious to us that for him Al Amarah was just some hell hole of a town he'd never heard of in the middle of nowhere. Compared with what he was used to doing, he must have been expecting to be bored stiff. When the call came in, there'd been no other volunteers to go south with him from the unit. Instead, a regular Royal Marine had been collared for the job of being his number two.

"The common conception in Baghdad is there's nothing going on here." Buzz explained.

"Is it now? Terrific."

Buzz had been brought in for a specific reason. Our longs had a range of up to 1,000 meters. Any target farther away than that, and we were just pissing in the wind. Literally, because the smallest gust would blow the round off trajectory at that distance.

The Enemy Out of Range

That put a lot of places the enemy loved to use out of our range. The bus depot on the north bank was 1,200 meters away, and the Yellow 3 junction right by the OMS [Office of Moqtada al-Sadr] building was at 1,700 meters. It was infuriating, because we could spot them running around up to no good but were powerless to stop them.

Someone in Abu Naji had good connections with his unit, and had put in a request for one of their sniper pairs because of the additional range of the weapons they use.

We explained the problem to Buzz.

"Okay, roger that. We'll see what we can do for you."

Up at Rooftop, they unpacked their kit. They had two grip bags with them. One was full of ammunition. Out of the second came some whopping great big sights and two pairs of ear protectors. That meant only one thing.

Buzz finally unsheathed his valise. And there it was. A .50 caliber Barrett sniper rifle.

Oh hell yes.

Barrett's Big Bad Sniper Rifle

I'd seen a Barrett before, but never fired one. It was known in the trade as the big bad mother of the whole sniper rifle family. Weighing a whopping thirty-plus pounds, it measures five feet from the end of its specially designed square-shaped stock to the tip of the thickly grooved muzzle.

It was designed by the Americans primarily for use on the battlefield to take out armored vehicles; drivers or the engine blocks, it did for both. It was also excellent for destroying enemy inside strong defensive positions such as sangars.

The weapon took rounds the same size as the Soviet-made DShK [Degtyarev–Shpagin krupnokalibernyy] heavy machine guns that had cut us up so badly on patrol with the OPTAG [operational and training advisory group] sergeants. The regular army uses .50 caliber Browning machine guns too, but only on a heavy tripod or welded to the roll bars of Land Rovers.

Buzz's Prized Toy

The Barrett has an accurate range of at least 2,000 meters, and sometimes farther still. Simply, the more gunpowder there is in the casing, the faster and farther the bullet will fly. You can only use them in a static position because it's too bloody heavy to carry around on patrol. The IRA had a Barrett in Northern Ireland and used to fire it from inside a specially modified car trunk. They wreaked havoc with it for a few years on isolated army patrols in bandit country.

Buzz's toy was going to do us a whole load of favors, and we were tickled to bits just at the very sight of the thing.

First, he put down a couple of zeroing shots into some rubble on the dam to make sure the journey hadn't screwed up his sight settings. That's when we really understood the need for the ear defenders. Hearing it fire was a joy in itself. It made a deafening boom like a miniature artillery piece, and gave off an echo that lasted a good ten seconds. A big puff of dust erupted from the sandbag wall beneath the thing, and the whole wooden sangar quaked on its foundations. From that moment onward, we dubbed it "the Beast."

"Damn, man, that thing's awesome," whispered Chris.

"Tell me about it. Imagine what one of them slugs would do to an OMS man's guts, eh?"

"What guts? Put a round through his kidneys and he could stick his hand through his body to wipe his butt."

A Technicolor Explosion of Blood and Flesh

Neither of us wanted to look unprofessional in front of Buzz and John, but it was bloody hard to conceal our excited giggling.

The pair didn't have to wait long for their first long-distance kill.

That afternoon, they spotted what must have been a senior OMS man standing on a rooftop right at the back of the bus depot. He was coordinating a group of gunmen having a hefty go at us from the north bank and an AK-47 was slung across the front of his body.

Sitting next to them, I followed the shoot through the sights of my L96.

The target was at least 1,600 meters away—the equivalent of sixteen soccer fields placed end to end. It was right on the limit of my eyesight in the heat mirage, even through my Schmidt & Bender sight. I had to strain to make the guy out.

Buzz fired. The round impacted right on the firing mechanism of his AK. Then a technicolor explosion of blood and flesh. Simultaneously, the man went flying backward out of his flipflops like he was a puppet on strings, and straight off the back of the other side of the roof. No more senior OMS man.

It was the first of a good handful of kills that afternoon. The insurgency's increasing mayhem provided no shortage of targets. Buzz was loving it.

"I thought it was supposed to be quiet down here. Is it like this all the time, Dan?"

"Yup."

A Mean New Beast in Town

"Bloody excellent. This is proper war fighting down here, you know?"

"Yes we do, mate."

"We don't get any of this sort of work in Baghdad. If only the guys knew what they're missing now. They'll be gutted when I tell them about it."

At midnight, after a good twelve-hour session, Buzz and John announced they were going to get their heads down for a bit. They'd been traveling overnight, and, amazingly, the Beast actually seemed to have quieted things down just a fraction in the city. There was one mean new bastard in town and they all knew it. The noise of the Beast alone was enough for the OMS brass to sit back and ask themselves what the hell we'd got our hands on. Then there was the damage it did to their men who'd got on the wrong end of it.

"I'll leave my rifle up here." said Buzz as he got up. "So any of your lads can use it if a long-range target pops up. Don't be shy with it, she's a real beauty."

No danger of that, matey.

"Use this bag of ammunition."

He chucked over a tightly cross-squared bag fastened by a draw cord. It reminded me of the old bags you used to keep your gym shoes in at school. We respectfully waited until they'd got at least as far as the stairs down into the house. Then, as soon as they were out of earshot, we were like little kids in a sweet shop.

"Oi, Danny, pass the Beast over here." whispered Smudge immediately. He was salivating to have his photo taken while firing it. "They did say don't be shy."

"Forget it, Smudger. I'm the boss here, and anyway—you're not qualified to use it."

A cheap trick, but true. And the only time in my whole career that I've ever relished quoting stupid army red tape. Since Chris was also qualified to use such a marvelous weapon as the Beast, I justified us a few practice shots just in case we did need to have a go at any long-range targets.

And marvelous she truly was.

Chris snipes and Des spots while concealed on a rooftop in downtown al-Amarah.

Full Metal Jacket and More

In every sense, the Beast gave off one hell of a kick. If you didn't grip it good and hard, it could recoil off your shoulder blade and smack you in the face hard enough to crack your jawbone. Also, you'd be deaf for ten minutes without the ear protectors. Chris and I put a couple of rounds into the giant metal leg supports of Yugoslav Bridge. They made a terrific row just rocketing around off its different struts.

While Smudge posed for his snaps, I had a poke around Buzz's gym bag. There were three different types of rounds in there: the normal "full metal jacket" brass ball rounds (like our green spot but a shit load bigger), some with tips painted yellow and red, and a third lot with tips painted gray.

"Right, give us the Beast back over here. I want to know what these yellow and red ones do."

I popped one into the chamber, pulled the cocking bolt back, and let it fly at the bridge again. On impact, it gave off a big bright yellow ball of light. Excellent. They were flash-tips, to illuminate the target so you could see where your rounds were hitting at very long distance.

Oost and Des were awe-struck, and watched every movement I made like two obedient little puppy dogs.

"Try a gray one, Danny." suggested Des.

Blowing Up an Insurgent Shelter

"Yeah, let's see what the gray ones do."

I popped a gray one off. *Boom*. It impacted on the bridge with a bloody great explosion.

Des was beside himself. "Wow, man! What the hell was that bad boy?"

I had a good idea. I took aim at a car that someone had abandoned on top of the bridge. It had been bothering us there anyway. OMS fighters could use it as cover to shoot at us. At least that was my excuse.

Boom!

The round piled straight through the engine casing and exploded somewhere in the middle of the block, causing a small fire to ignite. Yes. They were armor-piercing.

"Awesome, man, awesome!" They were Des's new favorites.

An armor-piercing round fired from a Barrett would punch through steel with some ease. It's greatly strengthened casing and specially shaped nose do the initial damage, before the bursting charge encased within its body finishes the job.

Chris and I popped a good dozen more grey rounds through the abandoned car until we found its petrol tank. Then it properly exploded and burnt down into just a shell we could easily see through. No more hiding behind that.

Buzz was back up at dawn.

"By the sound of things, you had a decent turn on the rifle last night, eh? I forgot to tell you, don't use the gray-tipped rounds. They're armor-piercing and they're really expensive."

Oops.

A Few Days of Death-by-Beast

"Ah, right. Sorry, Buzz, might be a little late for that . . ."

Buzz and John fitted in very well on the roof. By and large, they worked at their own pace and picked out their own opportunity targets. They didn't need me to spoonfeed them anything. That was their discipline and expertise, and it was fine by me.

Buzz didn't tell us much about what he did elsewhere. Out of respect, we didn't ask. It was enough for us just to know they were there with the Beast. They certainly made life a bit more difficult for the enemy's hordes. After a few days of death-by-Beast, the OMS coordinators learned to get their heads down and were forced to go about their warfare in subtler ways.

The Most Lethal Sniper?

So What Really Is the True Test of a Sniper?

By Craig Roberts

Roberts shooting M40A3 sniper rifle with Schmidt & Bender scope at the USMC Scout Sniper School at Camp Pendleton.

"In battle, the only shots that count are those that hit. . . ."

This quote is from a speech given by President Theodore Roosevelt at Haverhill, Massachusetts, on August 26, 1902. This quote is often used when discussing military marksmanship, but it leaves out the second half of the sentence: ". . . and marksmanship is a matter of long practice and intelligent reasoning."

Many excellent marksmen have exhibited superb skills with the rifle through the annals of American military history. From the American Revolution, through the Civil War, and on through all wars since, American riflemen have proven their skills in battle. Whether they be individual infantrymen like Sergeant Alvin York in World War I, or specialized sharpshooters of the Civil War era, or later snipers with specialized equipment, training and missions through the rest of America's wars and conflicts, many individuals have proven themselves with "shots that count."

Recently a book was published titled *American Sniper: The Autobiography of the Most Lethal Sniper in U.S. Military History.* It is the story of Chief Petty Officer Chris Kyle, a Navy SEAL who served four tours in Iraq as a sniper. It is a fascinating read, and Chief Kyle is truly an American hero and great warrior. However, when any claim to being the "most lethal sniper" is made on a book cover, it draws attention. The first question that enters one's mind is what makes someone the "most lethal sniper?"

In some circles, especially the general public, and too often book editors (who normally write the sub titles), the answer is simple: body count. But body count alone in the sniper community is only one factor to consider. "Body count" is like "counting coup." But does it really become the deciding factor on who is the best sniper in U.S. military history? The simple answer is no.

Roberts (right) with Lt. Col. Brown at SOF elk camp in Colorado. Roberts holds his Remington 700 sniper rifle mounted with Millett tactical mil-dot scope.

History of American Military Snipers

Comparing sniper activities in one era of U.S. history, or in one part of the world, with another is like comparing camels to water buffalos. One has to understand the history of American military snipers. When the U.S. finally began training and equipping specialized snipers in World War II, we had to start from scratch. We mounted telescopic sights on World War I–vintage 1903 Springfield rifles (and later the M1–D and M1–C Garand), then trained infantrymen who were good shots in the arts of camouflage, movement, long-range shooting, scouting, and reporting, then decided how and where to use them. Eventually snipers in both the Army in Europe and the Marines in the Pacific were recognized as having very special skills that were a huge asset to ground combat.

Roberts with M40A3 at the Scout Sniper School range at Camp Pendleton, California. Rifle is a vast improvement over the original M40 and incorporates a Schmidt & Bender scope, adjustable cheek piece, bipods, 10-round box magazine, and Picatinny rail for attaching a night vision scope.

But when the war ended, the sniper program went away. It simply fell through the cracks due to the failure by top leadership to appreciate the sniper as a combat multiplier. Then in Korea, we had to start the program all over again. We had the rifles and some of the old "re-tread" snipers from WWII, but had to train new men in the art and develop new tactics to address the mountainous terrain of Korea, and a large determined enemy that often outnumbered us. Our snipers did a great job, but then again, when the Korean War ended, so did the sniper programs and the rifles went into storage.

Ambushes, Booby Traps, and Sniping

In 1965 in Vietnam we encountered a determined guerrilla force that specialized in three things: ambushes, booby traps, and sniping. In the rice paddies south of Da Nang and in the Mekong Delta and everywhere in between, Viet Cong and later North Vietnamese Army snipers engaged our troops on every occasion where they felt they had an advantage.

Once again, we did not have a sniper program or a sniper school in either branch of the service. The military had once more been caught with its pants down, and we were facing snipers who had sniper rifles with telescopic sights and knew how to use camouflage, cover, and concealment very well. It was then decided that we needed to do something and fast.

For the Marines, hunting rifles were brought out of Special Services (the place where you checked out sporting equipment such as swim fins and volleyballs), and target rifles were sent over from Quantico to fill in the gap until something else could be developed. Our mainstay was the Winchester Model 70 in .30 caliber U.S. (.30-06), mounted with Unertl fixed 8X target scopes. Neither

Roberts with GySgt Carlos Hathcock, UsMC Ret., teaching a police sniper class to the Tulsa, Oklahoma, Special Operations team in 1989.

the rifle nor the optics were especially suitable for combat or the tropical climate of Vietnam, where temperatures ranged from 85°F to 130°F, and humidity could go up to 100 percent in the driving rains of monsoon season.

Still, Marine snipers like Sergeant Carlos Hathcock managed to learn the art of sniping in-country under the tutelage of Captain James Land with the 1st Marine Division and Captain Bob Russell with the 3rd Marine Division, and later managed to rack up huge body counts. Later, armed with the USMC-developed M40 Sniper Rifle (a Remington 700 with heavy barrel, oiled stock and Redfield 3x9 Accurange scope), Chuck Mawhinney followed suit and racked up an even higher body count. Hathcock is credited with 93 confirmed kills and 350-plus probables, Mawhinney 103 with 216 probables, and Eric England, "the Phantom of Phu Bai," with 98 confirmed and many more probables.

Down south, the Army fielded men like Staff Sergeant Adelbert Waldron who accumulated 109 confirmed kills with his M21 sniper rifle. SSG Waldron did have an advantage: He had the M21 semi-auto rifle with the Leatherwood ART scope. The M21 was an accurized M14 and the ART scope had ranging capabilities that compensated for bullet drop. The M21 proved much easier with which to engage multiple targets in a short period than the bolt-action rifles used by the Marines.

Major Powell's Sniper School

In 1968, Major Willis Powell established an Army Sniper School in Nam using the accurized M14s for the Army. The school lasted eighteen days and had a 50 percent pass rate. By December 1968, the Army had fielded seventy-two qualified snipers and the kills began to mount. In the period January–July 1969, Army snipers accounted for 1,245 kills.

After the Vietnam War ended and snipers in both branches had once again proven their worth, the Marine Corps and later the Army decided to establish a dedicated Scout Sniper School. The Marines set up their Scout Sniper Instructor School at Quantico to train instructors to go out to the divisions and teach at divisional schools (GySgt Hathcock was senior NCO at the Quantico school when it began), and the Army sent instructors through the Marine school, then established their own sniper training program at Fort Benning. This time, the sniper programs were not dropped by the military just because we didn't have a current war or conflict.

U.S. snipers became a very important part of all combat operations from Grenada, to Panama, Beirut, then Desert Shield/Storm and covert operations around the world. The sniper community, thanks to the above pioneers, was finally here to stay.

Long-Range Precision Marksmen

But it wasn't always this way with either the top brass or the civilian world when it came to appreciating our long range precision marksmen. Snipers in early years, including Vietnam, were considered as bushwhackers, back shooters, and even called by some as "murder incorporated."

I think this mentality really began to change not due to the skill of the men with rifles or feats of heroism, as much as two things that happened back in the late 1980s, when Charles "Bill" Henderson wrote *Marine Sniper: 93 Confirmed Kills* about Carlos Hathcock, and the following year publication of One Shot—*One Kill: America's Combat Snipers* by myself and Charles Sasser. After these two books became huge sellers, other books followed extolling the virtues of military snipers and their effect on the battle field. Movies followed, then more and more manufacturers began developing newer and better equipment for both military and police snipers. By the time we went into Iraq, we had great schools, wonderful equipment, superb training, and a whole new mindset about snipers in general.

Most Lethal Sniper Label not so Simple

Now to the point. It is impossible to make a solid statement that any one certain person is the "most lethal sniper in U.S. military history" unless one simply considers confirmed kills or body count as the sole indicator. There is much more to consider when one consid-

ers the lethality of any military sniper, anywhere, if one understands and takes into consideration all the factors involved. These factors must include history context, location, terrain, enemy, equipment, training, preparation and much more. For example, let us compare sniper activities in I Corps of Vietnam in the 1960s with Iraq in Chief Kyle's time.

Vietnam was not exactly a sniper's paradise. The terrain

varied greatly depending on what part of the country you were in. From jungles down south, to the Central Highlands, to rice paddies and villages near Da Nang to the mountainous regions of Quang Tri up north near the DMZ, every area had its own problems and demands. As an example, Carlos Hathcock worked the areas around Da Nang, which included steep mountains covered with vegetation to flat rice paddies that surrounded villages from which the enemy could hide in and attack at will. Most villages held tunnel complexes, and the enemy would often only come out at night. A sniper had to set up out in the "boonies" with his spotter, usually with no other friendlies around to lend support, and wait for the VC to start on their way home from attacking or ambushing marines on patrol. Distances of shots often ranged from only 200 or so meters out to 700 meters. The 1,000-meter shots were extremely difficult due to the heat, mirage, scope power, humidity, and bullet drop of a 173-grain .30 caliber round at that distance. We had no ballistics computers, no ranging reticules, no laser rangefinders, or any other space age devices. We relied on a map and compass and hand drawn range card. Our "ballistics computer" was a ranging card of the dope on our scope that we could use if we just didn't use "Kentucky elevation." Our scopes were simple crosshairs. (The later Redfield did arrive with a plastic ranging "tombstone," but it didn't last in the heat, so most Marine snipers simply cranked the scope up to 9 power and left it there.)

The Coriolis Effect

Sniper equipment improved and changed over the years. Other factors entered the picture for extremely long-range shots, but some are more impressive in terminology than in actual fact. One factor often mentioned these days is "Coriolis effect." Coriolis effect is a physical factor that influences the movement of spinning objects. We see mention of Coriolis effect on sniper shots having to do with the rotation of the earth and its influence on a long-range shot. It is very overrated, and the actual effect due to the earth's axis and spinning moment is negligible. A more correct factor would be what we used to call "spin drift." Spin drift is the Coriolis effect on the spinning bullet itself. A bullet eventually begins to drift in the direction of the rifling: fired from a right-hand twist, the bullet eventually will drift to the right, and from left-hand rifling, to the left. At 500 meters and beyond, it can definitely affect the point of impact. As my friend Tom "Moose" Ferran, Chairman of the U.S. Marine Corps Scout Sniper Association and a Vietnam war Marine sniper with forty-four confirmed kills, said, "the Coriolis effect is relative to ICBMs, airline travel, naval gun fire, and perhaps artillery fire as their flight times are long relative to small arms flight times. The projectiles from small arms are traveling at the same speed as the earth's rotation and so is your target. This Hollywood crap from the movie *Shooter* made me almost fall out of my chair when the earth's curvature was mentioned. The wind, the wind, the wind. Distance and altitude are mechanical calculations. The wind is human judgment unless you have one of those newfangled computer gadgets, hopefully without a dead battery."

Taking into account the curvature of the earth? Sniper fire is line-of-sight and the curvature of the earth has little effect on anything. If you can see it, you can hit it. The time of flight of a sniper round is too short to be dramatically influenced by the earth's curvature or rotation of our planet. For example, a .50 caliber 750-grain bullet's time-of-flight to 1,500 meters is only 2.35 seconds. Bullet drop is more of a factor in zero wind conditions, as the shooter must compensate for a 890-inch drop-to-target at that range. That's 74 feet! The next factor would be wind. Remember, a military sniper is trained to shoot center-of-mass of a human-size target, so any hit is a good hit if it kills or disables the target.

Iraq: A Sniper's Dream

Now to Iraq. The two main types of terrain are wide open desert and urban areas. Both are a sniper's dream. Wide open areas give you great fields of fire and engagement ranges. Urban areas provide you with good hides that provide both cover and concealment. Both types of terrain do not allow the enemy to stay as hidden as the areas where we operated in Vietnam. Unlike the Viet Cong, the jihadists didn't mind exposing themselves in many cases. In fact, in one three-day engagement in Baghdad, two U.S. Marine snipers, Sgt. Joshua Hamblin and Sgt. Owen Mulder, who were stationed on top of an Iraqi military facility, killed thirty-two Iraqi soldiers in the street to their front because "they just kept coming to work like Hussein was still in charge." They also destroyed several Fedayeen fighters and a truck full of RPGs from the same position. Hamblin later told me that their biggest enemy was fatigue. "We didn't sleep for three full days and lived on adrenalin!"

Any combat veteran, sniper or not, will tell you that in an engagement, everything depends on several factors, not just how many targets you destroyed. It involves the enemy's will to fight, their numbers, the terrain, his equipment, your equipment, motivation, morale, physical condition, training, leadership, and often a bit of luck. A target-rich environment doesn't hurt either.

Records Being Set and Broken

As for equipment, many records are being set and broken. Hathcock's .50 caliber shot of 2,090 meters (6,857 feet) with a Browning M2 machine gun set on single shot and mounting his Unertl scope was later broken by Canadian Corporal Rob Furlong shooting a .50 BMG McMillan TAC-50 rifle in 2002 in Afghanistan. His target was at 7,972 feet. Furlong's record was later bested by British Army Corporal Craig Harrison, who shot two Taliban machine gunners at 8,120 feet with an L115A3 .338 Lapua Magnum. Due to the range involved, would this last feat make Harrison the most lethal sniper in the world?

Combat sniper operations are not a contest. Every situation is different, as are the capabilities of the snipers involved due to equipment, terrain, and so on. But before we can label anyone "the most lethal" we have to also compare what they had to work with, where they were, and all the other factors discussed above.

One of the weapons Chief Kyle had to work with was a .300 Win-Mag sniper rifle mounted with a Nightforce 4.5–22 power scope. Compare that to what we had in the early days of Vietnam: a heavy barrel .30-06 target rifle with a simple 8X target scope with wire crosshairs. There is a definite advantage to better equipment, which can easily translate into a higher body count.

The True Test of a Sniper

The true test of a sniper is not just body count or long range shooting; it's whether your bullet hits the target, even when all your battery-'operated gear goes belly up. No computer, no range finder, no illuminated scope reticules. Just you, a map, a compass, a range card, and your crosshairs.

Subtitles aside, Chief Kyle is a true American hero and great warrior. So are all snipers, soldiers, marines, and other warriors who have served at the sharp end of the stick. To our enemies: Take heed, we are all lethal!

SEAL Chris Kyle: America's Most Lethal Sniper

By SEAL Chief Petty Officer Chris Kyle, USN

Charlie Platoon of SEAL Team Three during a deployment in Ramadi, Iraq.

While in Bagdad in 2002, SEAL Chris Kyle on one occasion had to deal with loonies who were attempting to infiltrate across a river from a marshy area. His team was located in a two-story house overlooking the river separating the city; the marsh was completely overgrown with weeds and brush. Kyle and company received probes from the area nightly and Kyle recollected, "Every night I would get my shots off, taking out one or two or sometimes more."

He went on to say that baddies would pop up from the brush, snap off a few shots, and drop out of sight. Even so, he dropped eighteen or nineteen in a week; the rest of the team raised the score to

The sniper hide Kyle used when covering the marines staging for the assault on Fallujah. Note field expedient firing support—a baby crib turned on its side.

thirty or more. Kyle describes one of the most bizarre attempts of the terrs to cross the river with beach balls—believe it or not! He also describes his his longest shot.

Beach Balls and Long Shots

I was watching from the roof one afternoon when a group of roughly sixteen fully armed insurgents emerged from cover. They were wearing full body armor and were heavily geared. (We found out later that they were Tunisians, apparently recruited by one of the militant groups to fight against Americans in Iraq.)

Not unusual at all, except for the fact that they were also carrying four very large and colorful beach balls. I couldn't really believe what I was seeing—they split up into groups and got into the water, four men per beach ball. Then, using the beach balls to keep them afloat, they began paddling across.

It was my job not to let that happen, but that didn't necessarily mean I had to shoot each one of them. Hell, I had to conserve ammo for future engagements. I shot the first beach ball. The four men began flailing for the other three balls.

A close-up of Kyle's .338 Lapua, the gun he made his longest kill with. You can see his "dope" card on the side, which contains the adjustments needed for long-range targets. It is a McMillan sniper rifle topped off with a 5.5–25 Leupold scope. His 2,100-yard shot exceeded the card's range, and he had to eyeball it.

Kyle leads a training session for Craft International, the company he formed after leaving the Navy. He notes, "We make our sessions as realistic as possible for the operators and law enforcement officers we teach."

Snap. I shot beach ball number two. It was kind of fun. Hell—it was a lot of fun. The insurgents were fighting among themselves, their ingenious plan to kill Americans now turned against them.

"Y'all gotta see this," I told the marines as I shot beach ball number three. They came over to the side of the roof and watched as the insurgents fought among themselves for the last beach ball. The ones who couldn't grab on promptly sank and drowned. I watched them fight for a while longer, then shot the last ball. The marines put the rest of the insurgents out of their misery.

Those were my strangest shots. My longest came around the same time. One day, a group of three insurgents appeared on the shore upriver, out of range at around 1,600 yards. (That's just under a mile.) A few had tried that before, standing there, knowing that we wouldn't shoot them, because they were so far away. Our rules of engagement allowed us to take them, but the distance was so great that it really didn't make sense to take a shot. Apparently realizing they were safe, they began mocking us like a bunch of juvenile delinquents.

The forward air controller came over and started laughing at me as I eyed them through the scope.

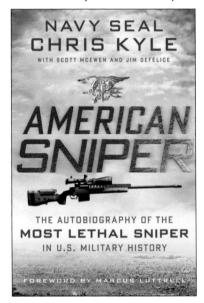

"Chris, you ain't never gonna reach them."

Well, I didn't say I was going to try, but his words made it seem like almost a challenge. Some of the other marines came over and told me more or less the same thing. Anytime someone tells me I can't do something, it gets me thinking I can do it. But 1,600 yards was so far away that my scope wouldn't even dial up the shooting solution. So I did a little mental calculation and adjusted my aim with the help of a tree behind one of the grinning insurgent idiots making fun of us.

I took the shot. The moon, Earth, and stars aligned. God blew on the bullet, and I gut-shot the jackass. His two buddies hauled ass out of there.

"Get 'em, get 'em!" yelled the marines. "Shoot 'em."

I guess at that point they thought I could hit anything under the sun. But the truth is, I'd been lucky as hell to hit the one I was aiming at; there was no way I was taking a shot at people who were running.

That would turn out to be one of my longest confirmed kills in Iraq.

Photo of Kyle sitting on chopper. Before his death, he offered a chopper training course for Craft. Note company logo and slogan in upper left-hand corner of photo which reads "despite what your mama told you…violence does solve problems."

One-Mile Kill Shot

By Staff Sgt. Steve Reichert, USMC (Ret.)

Bad Friday on Route Jackson

Eight years ago, I found myself participating in one of the fiercest battles ever fought inside the Triangle of Death. What started off as a normal squad patrol with fifteen Marines ended with hundreds of marines battling their way through a town. For years, I have been thinking about my actions on that day and what could have been done different. This is why I suggest everyone who carries a weapon for a living should bust their ass and train hard. You never know where you may end up and whose lives may be on the line. Below is my account from that day.

On April 5, 2004, Fox Company, 2nd Battalion, 2nd Marines participated in a five-day battalion operation to protect Shia pilgrims along Highway 8, aka Route Jackson. Over the course of the week, thousands of Iraqi Shia were expected to travel on foot between Baghdad and the cities of Najaf and Karbala in honor of the Shia religious event known as Arba'een. This was the first time in decades that the pilgrimage had been permitted and the number of people participating was greater than had been anticipated.

First platoon occupied a firm base in the Karch Oil Facility and the rest of the company occupied a firm base several miles to the south in an ice factory. The company sent out patrols throughout the day and night to deter attacks on the thousands of Iraqis walking southward. Fox Company positions were in Lutafiyah in the southern portion of the battalion's area of operations (AO), while the other two companies occupied positions to the north in Mahmudiyah.

The company experienced contact each day, either through the detection or detonation of improvised explosive devices (IEDs),

mortar attacks on the firm base, or short firefights with insurgents. The company firm base was attacked with mortar and machine gun fire each night and several marines were injured outside the ice factory when mortar rounds landed close to the company position. While the company's marines experienced multiple engagements throughout the week, no pilgrims were attacked within Fox Co's AO.

Tracers and Mortar Rounds Impact our Tank

I was the Platoon Sergeant for 1st Platoon at the time and was responsible for taking care of my marines and the platoon's operational control while advising the platoon commander. We had set up the patrol base the day prior in an abandoned building the locals turned had into a trash collection area.

It wasn't the best place, but it didn't stink and we could pull all the vehicles and marines under the shade. The night prior, the lieutenant and I climbed one of the highest tanks in the field located next to our patrol base. We did this at sunset so we could get eyes on patrol routes, but hopefully not skyline ourselves in the process. After we had figured out the route the 1st squad patrol was to take, we sat for a while and enjoyed the sunset. We had been up there now for about forty minutes when a long line of tracers appeared overhead, then started impacting the oil tank. Simultaneously, two mortar rounds impacted in the area. We quickly slid down the backside of the tank to take cover. The attack was over in twenty seconds, but drove home the fact that we were always being watched and the insurgents would exploit any opportunity we gave them.

Once we got off the tank, we launched a squad on patrol. Their first checkpoint was the area from which the insurgents had launched their attack. Then they were to head north while skirting the city on the east.

The squad didn't expect contact, since the insurgents typically didn't attack and stay in an area. The patrol was to return to base in four hours. As soon as the patrol reached their northernmost check-point, they were heavily engaged. Murphy's Law came into play and all communications with the squad went down. I remember looking to the north and seeing tracers flying through the air in all directions. I kind of felt like a father watching his kids get into their first fight in school and was hoping that they came out unscathed. As the other two squads were preparing to depart and link up with marines in the fight, communications were re established. No one was hurt and the squad was returning to base. When they had returned, I debriefed the squad. Plans for first light were to launch another squad into the same area to search for evidence left behind the previous night. You might not think that this would be important, but it really did help to know where your enemy had set up, how long they were waiting, if they were employing multiple belt-fed machine guns, bolt guns, and so on. Every bit of intelligence you could gather was important.

Observation Post: An Eye on the Town

The next morning, I had the marines assist in reinforcing an observation post on top of the tallest tank in the field. This would allow us to keep an eye on the patrols and, more important, an eye on the town. I took up an M82A3 (.50 cal sniper rifle) and an M40A3 (7.62mm sniper rifle) and, most important, a marine who could call wind. Cpl.

Tucker had been a range coach on Parris Island prior to getting to our unit. He had a great understanding of how wind affects bullets and how to read the wind. His ability to do so would prove critical that day.

The squad was to depart and head straight north to the southern edge of the town and then patrol through the easternmost part of the town until they reached the attack point from the previous night.

The patrol left the base around 0700 that morning. As soon they were moving toward the town, Cpl. Tucker, who was behind the spotting scope, noticed a dead animal on the side of the road. Not too out of the ordinary. However, he did pick up a slight reflection coming off the animal. I got behind the glass and noticed the same glint. The squad was notified of a possible IED and preceded with caution. As they approached the dead animal, they saw the wires leading into it so they set up a cordon per standing operating procedure.

Shortly after the cordon was set, we noticed women and children slowly clearing out of the area. All the marines on the ground knew what that meant, but were unsure where the attack might come from. Cpl. Tucker and I were actively scanning the area to their north for any sign of insurgent activity. Explosive ordnance disposal (EOD) was called and they gave a forty-five-minute estimated time of arrival (ETA) from their location. Moments later, a motorcycle with two men on it drove by the squad; in their passing they dropped two grenades. By the time the grenades had detonated, the bike had disappeared behind buildings and made its way out. Then it seemed like all hell broke loose and the squad started taking heavy fire from the northwest. I immediately called for our remaining two squads to mount up and to link up with the squad in contact and extract them. They were to come into the town via the main supply route (MSR), then head east to link up with the squad. After that call, both communication systems I had on the tower stopped working; we had zero ability to communicate with anyone.

The first shots out of my rifle were just over a thousand yards. Not an issue for the M82A3, especially since the one I had with me was one of the few that drilled. Cpl. Tucker and I were both lying on a metal roof, surrounded by metal barriers; this combination reflected a lot of sound and about made us both deaf after the first shot. We took the four seconds to jam in some ear protection; otherwise we would undoubtedly be completely deaf in minutes and practically useless.

Mixing Metal and Meat

The effect the Mk 211 rounds had on the insurgents was devastating. Each guy who caught one of my rounds was blown into pieces and left a pinkish mist in the air. The insurgents were trying to flank the

squad on the east, but our precision fires kept this from taking place. While all this was happening, the other two squads at the patrol base were loading up and getting ready to drive into the fight. Meanwhile, the company commander and 2nd Platoon left the Ice Factory to reinforce 1st Platoon.

The squad in contact was taking more and more fire as the minutes went on, so they started moving to find a defendable position. As they were moving north, two of the marines became separated from the squad and the insurgents tried to maneuver on them. However, Cpl. Tucker and I kept mixing metal and meat and the two marines were able to rejoin the squad.

The squad eventually found themselves deep in the town. In their search for a defendable position, they kept moving northwest. This made supporting them from my position even harder. As they would bound from one area to another, Cpl. Tucker and I would be scanning for targets of opportunity and killing them as they popped up.

Unlike how Hollywood might depict, it was not "one shot, one kill" 100 percent of the time. Most of the targets were moving and only stopped for seconds at a time. They were at various distances, and the winds were coming from multiple directions and gusting. If a round did miss the target and Cpl. Tucker saw the splash, he fed me instant corrections and the next round was out a second later. This enabled us to connect a number of times on distant targets under bad conditions.

Surrounded by Insurgents

The squad finally located a small schoolhouse they could easily defend. They cleared the school and set in for one hell of a fight. The school was the best defensible position around and they were surrounded by insurgents the second they took it over.

While the squad was defending their position, Cpl. Tucker and I kept the insurgents off the rooftops, for if they got access to a vantage point they could lay down effective machine gun fire and kill my marines. I was not about to let that happen. We noticed three

insurgents with belt-fed machine guns heading up a tall building north of the schoolhouse. The rooftop of that building would have allowed the insurgents to fire down into the schoolhouse and also keep reinforcements from reaching the schoolhouse. Taking this team out became our priority. The first round we sent out was off by a few mils. Cpl. Tucker picked up the splash and gave me a correction. The second round was also off, but a lot closer than the first. The third round landed on the stairway wall they were crouched behind. The backside of the wall turned red and we didn't see any activity from that point on. During the debrief at the COC the next day, the distance from our position to the insurgent machine gun team was 1,614 meters as measured by Falcon View.

We Were Sitting on Hundreds of Thousands of Gallons of Flammable Liquid

EOD arrived with one of the battalion's Combined Anti Armor Team sections. As they approached the squad's position, the CAAT section saw the muzzle blasts from our position and mistook those fires for an enemy sniper. They engaged our position with .50 cal machine gun fire as they attempted to link up with the squad. This went on for a while. We would fire, and then get a volley of .50 cal back at us from multiple machine guns. When the rounds started impacting the oil tank, I became a little worried. . . . After all, we were sitting on top of hundreds of thousands of gallons of flammable liquid. Lucky for us, the Russians who built the tanks built them well and they withstood the hundreds of rounds of .50 cal that slammed into the side.

Due to the massive amount of fire received on their planned route, the company had to find an alternate route to reach the squad. The company CO, the battalion forward air controller (FAC), and the 2nd platoon commander (along with a squad from 2nd platoon) were moving to establish eyes on the enemy when they were engaged from the east–west road on the south side of Hy Salaam and the palm grove. They returned fire and attempted to move to a position where they could better observe the enemy positions and call for air support. As they were moving farther east, I noticed an insurgent heading down an ally way with what looked like an RPK (light machine gun). As he turned, I noticed the fore end and the scope. By the time my round made it out there, he had jumped a wall and was gone. Seconds later Cpl. Speer (a squad leader in 2nd Platoon) was shot as he came around a corner. I did not know this at the time. By this time, the quick reaction force (QRF) from 1st Platoon had two casualties that needed to be airlifted out. I later found out that as the 1st Platoon QRF came under heavy fire, marines that I had once considered lazy, unmotivated, or weak minded were stepping up and taking charge.

They Were Bad and They Needed to Die

The battalion QRF was called in along with a Huey to evacuate the dead/wounded. When the BN QRF showed up with a full company (150-plus marines) and the BN jump command post (CP), both companies got on line to clear the town from south to north. Fighting was

intense for a few hours, but once the FAC got the F14s rolling, the insurgents started to head back home, and this was when time was on our side. From our point, we could see insurgents running back home with weapons in hand. Most would go inside and not come back out; those who did were now targets of opportunity. I remember one middle-aged man who ran to the hut in the back of his house. He dropped off his RPG and came back out with a pitchfork and began working his garden. Our first round out just missed him and he ducked behind a palm tree. Not only will the Mk 211 round penetrate armor, it will also penetrate palm trees! The second round made another pink mist cloud. These were bad people who had been trying to kill marines, so in my mind (and in accordance with the rules of engagement), they all needed to die.

As the day wound down, there were fewer and fewer people who needed killing. People started storing weapons and ammunition in coffins atop vehicles so they could drive them south out of the town past the marines. I saw one group dump a body out of a coffin and fill it back up with weapons. We were out of ammunition on Mk 211

by that time and using my M40 would have not been effective. Had we more rounds for the .50, I would have thoroughly enjoyed dropping more bodies around the coffin.

The sun was starting to set, so Cpl. Tucker and I packed up shop and got off the tower. When we got back to the patrol base, we were not surprised that everyone was gone. So we walked over to the guard building of the oil field. Inside, we found nine guards with AK-47s. We did our best to ask for a key to one of the trucks, and eventually Cpl. Tucker was able to talk the man out of the keys. Cpl. Tucker was a very diplomatic man who had certain abilities to persuade people. As we walked outside of the building, the armed guards followed us. They started looking around for other marines. The hair went up on the back of my neck and I started to talk on my non-operable radio. I then started pointing to various locations in the field and waving. The guards also started looking. I pointed to my sniper rifle and then pointed to more locations. I was trying to give the guards the impression that we were under observation from multiple sniper teams. It must have worked, because the guards got quiet and went back inside.

Two Men Kia: Haunted by the Thought of "What If"

As Cpl. Tucker and I were walking out to grab one of the vehicles and drive it back to base, one of the CAAT teams drove in to pick us up. As this was the last day of the Arba'een pilgrimage and the operation had concluded, the company then consolidated on Route Jackson and returned to Forward Operating Base (FOB) Mahmudiyah. Once we linked back up with the platoon, I learned of our two casualties and Cpl. Speer. When I heard what had happened to him

and where it had happened, I was saddened. All I could think of was the insurgent with the sniper rifle I saw and how he was able to kill Cpl. Speer before the other marines killed him. The ride back to the FOB was depressing. When we got back, I went to the company CP to pick up my platoon's mail. Next to 1st Platoon's mail was 2nd's. On top of 2nd's stack was a letter from Cpl. Speer's wife, with kisses and little hearts all over it. I knew she would be getting a visit by some marines in dress blues shortly and it's a visit no one wants to get. Could I have done something different and got the sniper before he shot Cpl. Speer? Maybe, but I'll never know.

It's one of those thoughts that will never leave you, always wondering "what if." Could I have done something different? Could I have trained harder or been faster on the trigger? Should I have had that area inside my field of view? I was surprised that no other marines were killed in that battle. It was twelve-plus hours of constant fighting with the sound of gunfire never leaving the town until sunset. This can probably be chalked up to the tactics the marines employed and the leadership of the commanders.

An IED Hit Changed My Career Path

Two months later, I was hit by an IED and was medically retired from the marines two years later. After that time in April 2004, I was training those marines in my charge as hard as possible. Even though my battle was over, I knew the training I provided the marines could possibly save their lives at some point and time. While I was going through the medical retirement process, I was reassigned to the Division Training Center (DTC). Upon arrival, the CO saw my drive and determination and asked me to stand up the 2nd Marine Division Pre-Sniper course. I couldn't have been happier; my CO was able to pull some of the Marines' most experienced snipers over to the DTC so they, too, could share their knowledge and experience before leaving the Marines.

After leaving the Marines in April 2006, I moonlighted at Blackwater, one of the best places to train at the time. After being there for a while, I thought it could be done better, so I started my own training company, Tier 1 Group (T1G). T1G has since trained thousands of Special Operations forces, training that has saved dozens of lives (this is documented). I left T1G a year ago to further improve on the way our military trains. If you think my last creation was good, wait until you see what I have in store.

An Ottoman Castle and a Syrian Sniper

By Chris Osmand

Chris Osmand standing next to a poster of OBL after his platoon raided a terrorist training camp in Afghanistan in January 2002. Osmand left the SEALs in 2006 and presently is the CEO of Tactical Assault Gear, www.tacticalassaultgear.com, (888) 899-1199.
(Photo courtesy of Author's collection)

Author Chris Osmand, is a former Marine and US Navy SEAL with multiple tours in Afghanistan and Iraq. Below is his description of the operations of SEAL sniper teams, which were tasked to provide security during the Iraqi elections in 2005.

SEAL Snipers Visits Tall Afar Election

The town of Tall Afar lies west of Mosul. The area commanders of Tall Afar wanted the Navy SEALs from teams Five and Eight to protect the voting booths for the upcoming Iraqi elections. The snipers had been so successful in Mosul that they hoped for a similar positive experience during one of the most important events in recent Iraqi history. Paul, one of my teammates, experienced the best two and a half weeks of his life in Tall Afar.

An American-held Iraqi hospital was one of the official voting stations. This particular area was a hotbed of insurgent activity and needed pacifying. The local commanders briefed the SEAL platoons on the insurgents' training, tactics, and procedures. The SEALs, now well-versed in their sniping operations, again used Strykers for the bait-and-switch operations that had served them so well earlier in Mosul. The snipers also set up unique shooting sites specific to the areas they had to cover. The overall battle space commander of Tall Afar had a great deal of confidence in the snipers and not only permitted them the use of the expensive Strykers, but in effect allowed them to do whatever they needed to accomplish the mission. It was of the utmost urgency to allow Iraqis the ability to live freely and vote during the election.

In the early hours of the combat operations, Paul witnessed one of the greatest shots he had ever seen taken by a fellow SEAL sniper. The "bad guys" were scouting the areas near the sniper hide sites but could not pinpoint any American positions. Paul mentioned that these insurgents did not wear traditional local Arab clothes and signal intelligence (SIGINT) indicated an impending attack. The three SEAL sniper teams were carefully monitoring the main supply route/road (MSR) when they spotted a man wearing a jacket way down the MSR.

A Gust of Wind and an AK

The SEALs knew from previous briefings that weapons were smuggled back and forth by individuals who hid them beneath their clothing. As luck had it, a wind picked up and revealed an AK underneath the man's jacket. The engagement began when one SEAL sniper fired one round that struck the man in his leg, putting him down on the street. After a couple of follow-up shots, the man was dead. No doubt the insurgents now knew that American snipers were in town. This

Navy SEAL platoon looking menacing at Tall Afar.
(Photo courtesy of Author's collection)

shooting incident triggered multiple well-coordinated engagements by the insurgents and Paul ranged his shooter at a target of 997 yards. Paul's sniper lit off a round and hit the Iraqi center chest and "laid him out like a fish." The amazing thing was that the SEAL shooter had not attended sniper school. However, he was an Olympic-caliber match .22 shot, and he was probably the best shot from the West Coast teams. Part of the SEALs' training also included an in-house, two-week, scout sniping program. SEAL Team-S conducted many such informal training courses in an effort to improve their tactics and procedures.

Dead SEALs Not Good for a Self-Serving CO

The multiple sniping engagements lasted for over two weeks until the election was finished. According to one SEAL, in Mosul and Tall Afar the SEALs had a 90–100 percent chance of contact when they left the gate. The four Purple Hearts earned by SEALs in this time proved how dangerous their work was. Finally, the commander of ST-3, Commander Riley, pulled them out because he was worried that dead SEALs would be bad for his record and promotion. Two American Army soldiers and three Iraqi soldiers had been killed near the old castle of Tall Afar.

One interesting point was that the insurgents soon realized the significance Americans placed on body counts, so they would do their best not to leave dead bodies in the streets. Many times women and children would appear and the body and gun, sometimes on a string, disappeared. If anyone picked up a gun, the SEALs shot them. Some Iraqis picked up flat cardboard and built a box around the weapon, which they then surreptitiously carried. Some Iraqi women also cleaned the area, as well, and the snipers saw them carry new dirt to the battle area to cover bloodstains. The SEALs sent Strykers to take a look at those areas and, indeed, they often found bloodstains

Navy SEAL platoon looking menacing at Tall Afar.
(Photo courtesy of Author's collection)

covered by new dirt. Other times some of those killed were not buried quickly. Those, the SEALs thought, were probably foreign fighters.

The Tall Afar castle provided a great sniping and counter-sniping position. It was so well-situated that the SEALs were able to hit any insurgent relief forces that wanted to rush into the areas the SEALs were covering for the election. Local Iraqi police and several Strykers supported the position. By 0300 hours, the SEALs had set up their fields of fire; the single best sniping positions were the lavatories used by the 82nd Airborne Division earlier in the war.

He Who Hesitates, Dies

On one day, Paul was bored and was throwing a turd at his buddy just as a coordinated attack was launched on the castle. More than a dozen mortar rounds, RPG rockets, small-arms fire, and carefully aimed sniper shots hit the castle. Coolly, the American snipers found their targets. Many insurgents were dressed in black with white headbands, making them easy to identity. Air assets tried to add to the insurgents' death toll but some of the close air support (CAS) was just not precise enough. Nevertheless, the firefight lasted six hours. The SEAL snipers learned a lot from trying to hit men sprinting at full speed at a distance of 600 to 1,100 yards. Paul estimated that the SEALs hit one out of every six runners. Any insurgent who hesitated usually got hit. "He who hesitates, dies."

One particularly tough nut to crack was a well-disciplined sniper, who turned out to be Syrian; the SEALs counter-sniped with him repeatedly. Counter-sniping required patience, a keen eye, and luck. It took three days to kill the Syrian sniper and it happened just as the SEALs' mission was about to end. The enemy sniper was so good that a mannequin the SEALs used to draw fire had three head shots. The Syrian used an HK sniping rifle, which the Americans ultimately recovered. Paul compared the sniping experiences to the German and Soviet WWII snipers locked in mortal combat depicted in the film *Enemy at the Gates*.

Paul tried to draw the Syrian's fire but could not get him to expose himself for even a split second. He was professional and disciplined, and throughout the day small-arms fire was intermixed with sniper rounds. So it was tough to isolate the enemy sharpshooter.

Bad Guys Sneak through Medieval Catacombs

Beneath the city of Tall Afar is a network of medieval catacombs. Paul spotted children playing soccer, occasionally moving through the underground cave network. Finally, the SEALs figured out that the insurgents were using the network to maneuver unseen into better positions.

The Syrian must have had spotters. The SEAL snipers thought that there was one expert sniper with a support team composed of probably four to five men who helped spot and snipe, as well. The Americans did have the advantage of superior technology. The Stryker had excellent optics to scan the areas, but the Syrian sniper was so good, he shot out the optics. The SEALs knew he was close, but could not get a precise location on him, only a general area. The snipers put out flank security to seal off particular areas.

SEALs used observation drills to locate the target. They scanned and shot anything suspicious, but nothing came of it and it was getting late in the day. Paul looked for an edge. The Stryker commander would not risk another expensive optical system, so Paul thought of using the Iraqi police as bait. He placed them on the top ring of the castle's battlements with orders for them just to shoot.

The day was ending. The mission had been a success. The Iraqis had voted, protected by a small but dedicated force of Americans. The SEALs started to wind down their operation. A few of their snipers were on the walls alongside the Iraqi police. Several men kept shooting, hoping to catch the Syrian sniper trying to get a hit. Maybe he would drop his cover, take a risk to get a kill.

Paul spotted an interpreter exposed at an archer window on the battlement. The SEAL told him to get away from the window, but shortly thereafter he got hit in the shoulder, the bullet striking an Iraqi police officer in the spine, as well. The SEALs medically evacuated the wounded out of the castle. Finally, one SEAL sniper from ST-8, using the observation drills learned in sniper school, spotted the small movement of a brick at a house closest to the castle, followed by a shot.

The Syrian Sniper Behind the False Wall

As more shots rang out, some of the SEALs assaulted the building but could not find the exact location. Then they found the place. It was a small, false wall hidden inside the building. The spotter would pick out a target and the Syrian sniper would move the brick, fire his weapon, and place the brick back into its hole. Although there was no body, it seemed likely that one of the American snipers had hit the Syrian, because the SEALs who assaulted the house found brain and other cerebral matter. Additionally and more importantly, they found the sniper's rifle in an alley as a car sped off.

The Tall Afar mission was a great success.

SECTION THREE: SNIPERS ON TODAY'S BATTLEFIELDS

8

BURMA

Ambushed In Burma

Report from SOF freelance writer, August 2010, down range in Burma

By Ian Maxwell Sterling

I have been bested now three days running by this cocky son-of-a-bitch, who eludes my every scheme and thrust for his scrawny, miserable carcass. The camp rooster—now my enemy for what remains of his short life—wakes me at zero-dark-thirty in a non-stop four-hour barrage. I mean to have him yet for my meal, but for the moment, I am the new meat here after having been gone for over twenty years. At that time (1988), I was with battered student revolutionaries after they beat a hasty retreat from the streets of Rangoon to seek refugee and solidarity with ethnic resistance forces in the deep jungles of eastern Burma. What I now witness in these same jungles is nothing short of astounding—a total surprise. Ambushed!

The World's Longest Running Insurgency Rages

I am mindful that *SOF* magazine had been "stung" before in this same region, where the world's longest-running insurgencies have raged, flagged and flashed again in proud moments since 1949. It was in in the early 1980s, as recounted by Bob Brown in *SOF*'s 25th anniversary edition, that he and his mates were taken (and also "got taken") in search of MIAs in Laos, only to be left high, dry, and well-fleeced for their passion and purpose. This is a land of sophisticated deceptions veiling ever deeper layers of illusion—the art of ancient ethnic civilizations and cunning cutthroats of all sorts bred in the daily battle for survival in what some call the Land of Sighs.

Today's ambush? I had gotten accustomed to the notion of highly factional ethnic hill tribesmen in ragtag home defense militias fighting a losing battle against one of the largest armies in Southeast Asia. I had also labored under the impression of student protesters in the streets of Rangoon and Mandalay taking on the Burmese Army frontally and bloodily. Back in 1988, it soon became apparent to me that despite agreement on their common enemy—the Burmese government and its army—the factionalism among ethnics, as well as the highly independent nature of the urban activists, hobbled any chance at effective unity of effort for the pro-democracy movement. The dictatorship has taken masterful advantage of this and further created rifts within and between each group through a campaign of intimidation, coercion, imprisonment, and terror. After two decades, the ethnic and activist resistance was reputed to be as disjointed as ever, fighting separate wars of survival against hopeless odds. With 60 percent of Burma's army concentrated in eastern Burma alone, ethnic fighters there would not seem to have much of a chance. The hopes of the urban revolutionaries were dashed when the international community twice failed to come to their rescue during the Buddhist monk uprising in 2007 and Cyclone Nargis in 2008. So I figured things were as bad as ever for the militant pro-democracy movement in Burma. Was I ever wrong! Down, yes, but definitely not out.

As I moved into this base camp in a hiding area well inside Burma, I started to notice things I had never seen before. The first thing was a wide range of uniforms . . . not because they were having to make do, but because I was seeing officers, NCOs, and soldiers from many different ethnic groups. As Custer found out the hard way about "massed Indians" with an attitude, so also I am witnessing a "massing of the clans" with a lot more than shared "attitude." With Burmese as their common language and the murder of their families and friends by the Burmese Army, this robust gathering was soon to reveal even more things in common.

This place looks the same as other encampments I have been to before, but something is different. I see it now in small symbols, in material never seen here before and in new ideas on planning. I noted in the latrine off to my right this morning, complete with concrete floor and local porcelain fixture, these words fingered in concrete by the soldier whose shit detail it was to build another shitter: "Fight for Freedom!" I wonder how many Burmese Army intel analysts grasp the significance of this obscure hillside graffiti in a forgotten bit of jungle in the middle of nowhere. One lone soldier. One voice. One message. There is awesome power in this spirit, beyond the enemy's dull imaginings. Simple words scrawled here are born of personal witness to the slaughter of innocent farming villagers by the Burmese Army and the brutal government terrorism against its political prisoners. The Burmese dictator and his army have stirred an awesome adversary in the will of these people—a people who no longer talk about just surviving and waiting for America to come to their rescue.

Decades-Long Scorched Earth Campaign

The State Peace and Development Council (SPDC)—the euphemism

for the dictator's repressive government—has waged a decades-long scorched earth campaign that has borne much fruit in terms of incredible profits from tribal lands seized—lands rich in oil, natural gas, precious gems, gold, uranium, and hydro power. SPDC's campaign, in its brutality in using murder, rape and slave labor as weapons, has, however, also reaped a dark bounty. It has ignited a massive fire of the Spirit now seething. People are talking about winning. Winning now! An old engineer friend interrupts my thought as he walks by at sunset carrying a huge .50 caliber sniper rifle built in one of a number of jungle machine shops. It looks like something out of Star Wars. "I have nine more bolts and barrels to work with now." He laughs and disappears, clambering to higher ground. Later I hear echoing booms as he tests out a new muzzle brake from American friends. It is pitch black and so I wonder what kind of scope he is using.

This place is stunning in its beauty and grandeur. Grunts of fortune have come here by ones and twos . . . sometimes even by the dozen. But except in the case of French mercs in the early '90s who established a commando training program (well done), most have contributed not much to these hill tribesmen of great pride, equal stubbornness and little means. Burma's ethnics have many reputations—one is as proud mates of Brits and Yanks during WWII in the BTO (Burma Theater of Operations—CBI [China, Burma, India] to you Yanks). It was here that these primitive men with their farming families proved their reputation as the best jungle mountain fighters in the world. Their deaths by the thousands, mostly at the hands of Japan's Burmese-led puppet government, came mostly in the form of villagers targeted for genocide by age-old ethnic Burmese rivals. The slaughter even horrified the Japanese by their own accounts. Bottom line: In this great war for liberty that we Brits and Yanks gained at our hill tribe buddies' expenses—liberty that these freedom fighters did not get and still pursue—the hatred between non-Burmese ethnic minorities and the Burmese majority was forever forged. It smolders and flames today just beyond my bamboo hut in a silent war in the shadows of exotic Burma. Some call it "a secret genocide."

I had been among student activists and ethnic groups of Burma for a season in the late '80s, when the hopes of democratic reform were at first bright, and then quickly dashed in a brutal put-down by the dictatorship. Given the lack of firepower by student activists, the reality in Burma then, as now, is that ethnic resistance forces are the only credible threat to the dictator's power. There is nothing quite like tens of thousands of freedom fighters, who have witnessed the killing of their families, as the driving factor to be reckoned with here. I was impressed by the ethnics' courage, but discouraged by their sometimes passive mix of Buddhist–Christian tolerance of the terror tactics of the Burmese Army, known by villagers and soldiers simply as SPDC—the kindly sounding name of this utterly ruthless dictatorship. I had found ethnic political leaders indecisive in their complex maneuverings and meanderings as they dallied around trying to talk their way to peace with the Burmese dictator, General Than Shwe. Not much was accomplished, at least through Western eyes.

A Silent Genocide

I shuddered at this notion of one's very own leaders betraying their people—a lot of that going on these days, and not just in Southeast Asia. This reminded me at once of America's hard experience

Fifty-caliber, scoped, home-grown sniper rifle packs a mule-like kick. It has been raining death and destruction on the Burmese sick thugs.

in Vietnam, where the imperative was "Don't lose to the Commies," another double negative without vision that brought us tens of thousands of brave young soldiers lost for all eternity. With all this hauntingly in mind, I had become discouraged after a while, left the hinterlands of Burma and drifted. I decided to come back only when I heard there was something anew in the wind. "Ethnics coming together and going hi-tech" was enticing first bait. I was hooked, but also doubtful. I proceeded back to old haunts along familiar routes from my past, trying to avoid old missteps and land mines of both the field expedient and cultural variety.

I noticed a few images that seemed totally out of place. Young, clean-cut faces, bantering away on satellite phones, sporting top-of-the-line ruggedized laptop computers in Pelican cases, and watching a You Tube video from a satellite Internet feed onto a plasma screen. "Want a no-foam latte?" Underground activists from Rangoon. These are the younger brothers of the revolutionaries I first met back in '88. A different breed now. Their older siblings had been a hearty lot taking on the Burmese Army and police boldly and dying by the thousands in bloody street fights. These guys are also bad, but in a high-tech sort of way. They bring an added dimension to the "resistance team" now unfolding before my eyes, causing me to reassess old concepts about what was actually going on under triple canopy jungle and underground in Burma's urban centers in this "gathering storm."

There had always been this hand-to-mouth method of insurgency here on-the-cheap—forever scrapping and somehow surviving until the saving rains of yet another monsoon season kept the Burmese Army from massing for attacks. This foot race against the clock has always been the "Ethnic Olympiad"—a run-for-your-life contest, in which the prize was life itself. But ethnics, who would portray themselves as plodding outback farmers, have just revealed a startling new side—as cunning, plotting networkers of the twenty-first century. These tenacious freedom fighters have

been quietly modernizing and networking in a manner that even the world's top trans-national threats might envy. Urban activists operating in parallel have evolved over the past twenty years and have emerged here in the night to explore new partnerships with old allies from '88—some of whom have been waging insurgency since 1949. Asphalt jungle revolutionaries join triple canopy jungle guerrilla fighters in common cause, since the dictator's stated aim is to wipe them all out in order to secure himself against opposition. Talk about clarity of purpose! Got it.

A Jungle Feast

"Home" here is a small encampment of ten bamboo huts perched on cliffs with stream beds cutting through separate towers connected by bamboo bridges. The boys have just brought in tonight's supper in the form of a dazed 4½-foot long lizard trussed up and slung over a shoulder, AK-47 on the other shoulder. The norm here, however, is old M16s for fighting (because AK ammo is rare and expensive). For chow, it is canned sardines in tomato sauce, routinely punctuated by chicken and goat, and infrequently by pork. Greens from the jungle of all sorts and without names are daily fare. One is a finely chopped course weed that tastes almost exactly like bacon, eggs, and green weed, a favorite among Westerners here. Eggs mixed with wild onions are SOP and dropped in huge woks of boiling oil, resulting in deep fried omelets that would make the stoutest heart surgeon get out the paddles. The mountain lads here (mostly in their twenties) have about 1 percent body fat—testament to life "on the run" in the toughest jungle mountains in the world.

Out-Numbered and Out-Gunned, Not Out-Run

These soldiers are out-numbered and out-gunned, but never out-run. They hit hard and fast in vicious contacts that span only a few seconds, then are gone in a flash, sprinting downhill like scalded cats barely touching the jungle floor. These jungle mountains breed simple, durable strengths of character and body. They also provide the defender an enviable terrain advantage. These guys kill their enemy at 100:1 ratios, and WIA ratios are even better. The simple facts are that out of necessity these farmers-turned-guerrillas are killers and good at their craft. These ratios are also testament to brilliant command at operational and tactical levels. This is all done on a meager diet of rice, oil, fish paste, chili, and the odd lizard that crosses their path. Tonight is movie night at my hooch. The unanimous pick is *300*. The boys are familiar with Spartan odds.

Movement around Burma comes in various forms of humping the boonies on harsh land routes. You are either breaking your ankles running streambeds or careening down black diamond slopes on mud. The locals will say: "You have to read the mud," but a stout bamboo walking staff is fair compensation for me, ever the novice downhill mud racer. At some point, you have to swim or boat your way across rivers. It can be on a bamboo raft of a mere five lashed poles with some skinny old fellow polling hard against the current, giving up speed for persistence and eventually making landfall well downstream. A rubber dinghy adrift moving downstream is another method, though infrequent and risky. Rivers are commerce-bearing life-givers for friend and foe alike, but rainy season can be a killer with swift currents and debris propelled like torpedoes by highland torrents. Nothing like a heavy teak log harpooning your ass broad-side in the middle of a starless night.

Chugging, long-shaft engine-driven long boats are the norm for locals, coming in all sizes from 2–3 man skiffs for scooting about to large "buffalo boats" able to move a hundred souls at a pop. In a very steep, dry streambed-turned-mobility route requiring me to work hand-over-hand, I step in more elephant shit than a man ought to, but I've never yet seen one of the beasts. In Bernard Fall's *Hell in a Very Small Place,* recounting the fall of Dien Bien Phu, one of the clear images is of giant staircases carved out of mountain sides for elephant convoys bringing in heavy artillery to encircle the French garrison. I pause and reflect on the determination of the Viet Cong long ago as inspiration now for tasks before these ethnics of Burma, and then I move on, minding my step.

Ambushed in Burma, Part II

By Ian Maxwell Sterling

*Interpreters are vital to the success of training. Good English speakers are few.
"Irregular warfare" once was interpreted as "immoral warfare," which caused
about a week's delay in training when ethnic leaders got up and walked out.*

An Organized Band of Brothers

Commanders and civilian leaders have gathered here and are using
a strange new language. I hear the terms "center of gravity," "decisive
points," "operational art," "sensor–shooter synchronization," "decision support template" and more. They keep me at a distance and
it is clear I am not welcome into their inner circle. These guys have
had a steady professionalization program "in the shadows" ongoing
since 2005! A private American–Brit–Canadian–Australian effort in
the main. This is not a loose bunch of thrill seekers, as is the norm
here, but instead an organized band of brothers, mostly with *SOF*
backgrounds. They come and go here, disappearing into the thick

darkness. All are volunteers and will not divulge their real jobs or pasts. Ranks vary from buck sergeant to sergeant major to warrant officer to officers up to the rank of colonel. Did I mention SWAT team leader and cyber hacker? They are tight-lipped to a man and give me precious little. The Americans range in experience from long ago elsewhere in Southeast Asia to just fresh from Southwest Asia. The others, to include a few swarthy-looking Chetnik killer dudes from Eastern Europe, reveal little, except for implying contract work in various hell holes in Africa. What is clear is that their efforts here have the ethnics talking and thinking in very different terms than ever before. It is pretty clear that if I am going to learn anything, I am going to have to get if from the ethnics.

My stature as a not-quite-outsider who has been here before has been raised a couple of notches, because of my bringing along a butt load of special technology as peace offerings to my hosts. In honor of this, I have the preferred seat at the head of the table and have been ceremoniously given the prized piece of meat for dinner. Should have seen this coming. I am confronted with what looks like the Full Monty appendage of a very well-hung water buffalo, complete with accompanying hardware. I manage to choke down this beast amid no shortage of digital camera flashes and roaring laughter. I was either being honored or suckered, and am still not sure about it all. At any rate, for my efforts I am rewarded with information. Fair trade, but I still hope those digital photos don't get out on the Internet. The possible captions keep running through my mind, none good. "Long dong on the Mekong. . ." Don't go there.

Expertise Plus Passion a Dangerous Mix

There has been a rather concerted effort running international mobile trainers across the region, as well as in urban areas. There has always been no shortage of loose cannon, Rambo wanna-bes running about in the Burma region conducting various training of equally varying quality, but this effort appears to be different. My hosts say that some round-eye millionaire came along back in 2006–2007 and then others joined in during Cyclone Nargis in 2008, providing a needed boost to ethnics, who were already expanding their "transformation" effort beginning back in the 2005 time frame. These backers basically supported experts in unconventional warfare to come into the region and work on professionalization of underground activists and ethnics. Some of this was done well down range, some through Internet methods, while some of this effort involved train-the-trainer programs in North America, in the case of information technology and technical skills. Expertise plus passion is a dangerous cocktail, indeed, for the dictator.

SATCOM Connection

The provision of extensive satellite communication systems for almost a dozen of the major resistance armies was one of many initial infusions into ethnic ranks with the clear intention of promoting collaboration and unity of effort. This I took to be a conscious effort to start at the top by connecting ethnic leaders in the field. Satellite phones and high-data rate BGAN satellite broadband were also provided to a broad range of activists I had met only a few months earlier in Rangoon, Mandalay, and one other city. Over a period of weeks I had met with an interesting mix of intellectual elites, underground leaders and independent student activists—each with unique perspectives and agendas. I did not understand the implications of this "SATCOM connection" at the time, but now it was becoming clear to me. There were little tell-tale indicators in the terms they used that I did not pick up on at first. But in retrospect, with all that I was now learning, it was all too apparent that they had been privy to the same trainers I was seeing here "in the woods." Follow-on training of commanders, underground leaders and key staff in the fields of strategic and operational planning, as well as the conduct of planning workshops and war-gaming have produced a new level

These are the final ten snipers who graduated. They are running their own sniper programs now.

of sophistication in the militant pro-democracy camp. These boys are good.

Day One was a Frikkin' Disaster

Tactical training has quietly focused on developing a cadre of trainers in critical skills, who then go off and train more trainers and then operators. I was told that one class of fifty sniper trainees was reduced to ten graduates in the end. Most of the international instructors were fresh from combat tours in Southwest Asia. They chose to have their trainees start training using their worst weapons—mostly Vietnam-era AR-15s and M16s and a couple of converted M60 machine gun barrels, the logic here being that the graduates needed to be able to work with whatever weapons were on hand in remote villages. As one of the instructors confided: "Day One was a total frikkin' disaster, man. I mean I had this goofy interpreter who totally sucked and then our first test fires had shot groups that looked like shotgun blasts at 200 meters. I asked what was the longest shot these guys would dare to take with their weapons. They all agreed that it was not more than 100 meters. I thought this was going to be hopeless. Then one of our SF commo men brought out some thick flex cuffs. These did quite well in tightening up M16 upper and lower receivers. Shot groups immediately tightened, morale soared, and my give-a-shit factor recovered a full notch. I got rid of the interpreter, too, so regained another notch. The ethnics brought in this new 'terp,' who rocked, and so we were soon off to the races."

Fifty candidates were cut down to twenty-four after barely a week. Oppressive direct sunlight smoked the less motivated ones who were used to the recesses of deep jungles. The urban activist guys were the first to fold. The tough hung in there, though, even with their sporadic shot groups. We finally had to resort to using Chinese-made .177 caliber air rifles with Bushnell 6X scopes to determine what the problem was with promising shooters. Our instructors tested these out and shot groups were so tight, we thought maybe these guerrillas needed to convert to poison-tipped air guns pellets, instead of mucking around with M16s. The slow muzzle velocity of the air gun was unforgiving in revealing trigger-control problems, and so we quickly were able to correct this problem, and at almost no cost. "Made in China." Bummer. We hate these guys. When the trainees were not on the firing line with M16s, they were off to the side "blazing away" one shot at a time with air guns. *Bam, Bam, Bam… ping!* Even inside our jungle classroom, we had the boys firing in lanes on both flanks throughout instruction as we taught in modules, sometimes over the heads of instructors.

Getting Ready for "Soft-Skinned Targets"

Trainers, students, and shooters alike soon learned to trust and concentrate no matter the distractions. After several weeks, we were down to but ten shooters, all now doing rather well at 300 meters. They loved the Millet fixed 10X scopes and quickly became proficient with Leupold spotting scopes and laser range finders. Then a brand spanking new black .308 with Leupold fixed 10X scope was introduced by one ethnic commander to see what they could really do. They were so fired up to prove their skill that they asked for a day off to clear off a whole hillside in order to take 500–800 meter shots. We had our doubts, but relented, and soon they were feverishly hacking down teak trees. Expensive bush whacking, for sure, but nothing quite like a sniper with an attitude. The "black gun," as the ten survivors called it, was a dream, as they all drilled tight shot groups again and again.

Upon my finally getting the green light for an audience with commanders, they admitted they did not like killing enemy soldiers, many of whom are just poor conscripts. The conscript-heavy forward battalions are attacking villagers and burning their homes—over 3,500 villages burned and mined in eastern Burma by latest tally with over 470,000 persons displaced. "We realize this war cannot be won by killing soldiers alone. There are other targets in our sights and hearts." Toward the end of training, out of the jungle, unexpected to us were produced several .50 caliber sniper rifles of the homemade variety. Incredible feats of jungle engineering, these beasts weighed a ton and kicked like mules, but found their marks again and again. When we asked our friends what they had in store for these systems, they only smiled and said that they had some very specific "soft-skinned targets" in mind. I should mention when I later visited their workshop, I noticed on a bench a technical manual for a larger caliber sniper system. It looked like either a 20mm or 25mm design, but the old engineer quickly scarfed it up and hid it from my sight. There were some huge empty shell casings on the floor, much larger than .50 caliber. Clearly my friends have soft targets in mind at greater range. More bad news for the bad guys, because the good guys can definitely shoot. *Ping, Ping, Ping* on the cheap…then *BAM* for the payoff!

The Jungle is Wired

One evening my new commander friends invited me their hooch for a special movie. Again I am given the seat of honor, but spared

the previous ritual repast, and the video rolls. It is of me on multiple trails at night taken over the period of several months since I
arrived. I am intrigued to say the least. It turns out these guys have
their jungle "wired." They have acquired remote-controlled, infrared
video camera technology from a corporate backer in the States and
have been installing it and recording the bad guys for several years.
Burma is a testing ground of sorts for this. When I asked why they
had not revealed their videos to the international community, I was
surprised at the answer. "The international community is asleep. Our
people die and politicians just talk and do nothing about it. We have
been sending photos and reports for years. After international community governments failed to get rid of the dictator during Cyclone
Nargis, we realized we are totally on our own. We realized we need
to do something new. So since 2008 we have been busy capturing
the Burmese Army doing bad things to our people on video. We are
working with an international investigation initiative. We think 'live
video truth' about what the dictator and his army actually do to our
people is the best weapon we have. Photos of what they have done
are not enough. It takes live video evidence. We call it 'video ambush.'
We cannot slowly leak video and get the same results. We know we
have to do one big ambush. We have been struggling like our fathers
and their fathers for many years. So we are patient."

They next talked vaguely about international backers that have
come on the scene recently. This includes Western corporate inves-
tors, who understand that ethnic lands are the richest in natural
resources. At some point these businessmen realize they are going
to have to cut deals with ethnics, and so this is being explored
with newfound interest. I try to pump them for more specifics,
but they evade this and instead suggest I visit the electronics
shop not too far distant. More perilous night navigation on mud
slopes, but worth it as I soon discover. Holy shit! RadioShack on
steroids. These guys are doing stuff I had not imagined. Somehow
the memory of digital cameras flashing at my memorable dinner now fades as I contemplate the implications of these cunning

warriors' stealthy new methods, networking and innovating with their backs against the wall. With bad guys closing in, there is clearly nothing to lose now.

As I walk back to my hooch, my middle-aged interpreter—a former front-line soldier and now my personal confidant—talks about all sorts of things that he thinks I should know more about. "The leaders now trust you more, because you listen and bring them the special hi-tech gifts. Good. You need to slowly ask them about the big raid, our SEALs, anti-aircraft technology, and other things. Do not go too fast. It is not our way. I will tell you when the time is right. You must not ask them direct questions. I will teach you a better way. My advice is you have much to learn and first need know how to 'read the mud' and quit falling down all the time." And then he bows and disappears in the night.

Electronic Jungle

As things wind down here, the urban dudes are playing video games in the hooch next door and laughing in high pitch; commanders huddle in hushed tones chewing beetle nut around a waning fire; young officers are clustered and blazing away on the Internet SATCOM link with a swarm of bugs swirling around bright fluorescent lights; NCOs are massed over in the movie hooch watching the final scenes from *Apocalypto*, and my three batmen have shown up as usual for evening tea and biscuits in ceremonious British tradition before I turn in for the night. Then that son-of-a-bitchin cock rooster takes his perch yet again, ever beyond my grasp. I am in awe of what I have witnessed here. I am left with much to think on and write about in days to come. I wonder if Americans understand that the spirit of 1776 thrives in the hearts of brave warriors—our old WWII allies from the past—now in their last stand. Whom better than these stout hearts should we be supporting, as they face the incredible odds now massing around them? Dawn will come early and opportunity knocks in the midst of adversity.

American Sniper in Burma

By Scott Johnson

Top: Colonel Ner dah Mya of the Karen National Liberation Army
with KNU soldiers. Jack Slade (an alias) is standing on the
right.
(Photo provided by Scott Johnson)

The war in Karen State has raged for decades, making it currently
the world's longest running conflict. It was in 1949 when the fighting began and ever since the Karen National Union (KNU) has been
defending themselves against the oppressive Burmese junta, ironi-
cally called the State Peace and Development Council (SPDC).

On January 31, 2011, I attended a ceremony at a jungle base camp
buried in Karen State, where Karen soldiers and political and religious leaders gathered to mark the sixty-second year of their strug-
gle for freedom. Here, the Karens moved freely and it was clear that
while the SPDC may control the urban areas in Burma, the countryside in Karen State is largely under control of the KNU's defense
force, the Karen National Liberation Army (KNLA).

Colonel Ner Dah Mya, a KNLA commander, organised the ceremony in what he called an act of "national pride." "Even though we are outnumbered," he told me, "we will show the Burmese regime that we cannot be defeated."

Attending the ceremony was a Buddhist monk nicknamed Rambo, whose presence signified the recent realignment of Christian Karen and Buddhist factions who after fifteen years of fighting each other were now joining forces against the junta. The monk gave a speech describing how the Burmese government once offered a five thousand Kyat (Burmese currency) reward for his head.

After the ceremony I was invited to accompany a KNLA patrol that was going deeper into Karen State. The patrol was going to resupply a unit involved in rebuilding a tribal village that was destroyed by the SPDC in late 2009.

The KNU/KNLA have commenced a village defence project involving rebuilding burned villages so refugees can reclaim their livelihood. Hundreds of Karen refugees had so far been resettled back from Thailand, but the danger remains nevertheless, as SPDC military camps were only kilometers away. One nearby camp was actually a captured Karen village where the SPDC had put their bunkers under the villager's houses to deter attacks from the KNLA. Such is the way of the brutal Burmese army.

It was on this patrol that I had the opportunity to interview a foreign fighter, a Westerner using the *nom de guerre* of "Jack Slade." He had extensive combat experience from conflicts around the globe and was happy to lend his expertise to the Karens. He had, in fact, killed two SPDC soldiers as retribution for burning a Karen village.

Our patrol set out in the morning led by Colonel Ner Dah Mya. Colonel Mya was the son of the legendary KNU General Bo Mya and, like his father, he was determined to live his life defending his people. All told we were a dozen or so armed Karen soldiers and two foreigners, Jack Slade and me.

The following are transcripts of interviews with Jack Slade, details confirmed by Colonel Mya and other soldiers on our patrol. In his own words Jack describes how he "terminated two enemy combatants as retribution for the destruction of a peaceful village." The sniping operation occurred in early 2010.

We came upon these Karen internally displaced people living in the jungle. Their village was destroyed by the Burmese army and the KNU/KNLA is building them a new village.
(Photo provided by Scott Johnson)

Retribution

Jack Slade: We had 150 SPDC troops coming at us from one way and 150 Democratic Karen Buddhist Army (DKBA) that were allied to the enemy coming at us from another. They actually crossed into Thailand to sneak up behind us. They came by trucks from the nearest SPDC camp during the early morning hours, then hiked the last couple of miles on foot. The KNLA guys helped pack up all of the old, crippled, or pregnant women villagers and their children in the KNLA trucks and drove them into Thailand. Then the soldiers came back and tried to get all of the men and women that couldn't fit in the trucks to walk the three miles over the border as fast as they could.

It was decided the KNLA troops, me, and one of my teammates would try to draw the SPDC/DKBA's attention away from the village. We circled around to the hills farther into Burma just as the SPDC and DKBA opened fire with machine guns and mortars on the vacant village. We sent a Karen sniper team around to fire at the mortar men, but after a few shots they had to abandon their position due to a large volume of machine gun fire shredding the trees around them. The SPDC then divided into smaller eighty-man groups, combined with the DKBA, and began to give chase.

We played ambush and counter ambush for twenty-two days, walking eleven days into Burma then eleven days back. We foraged for our food and by the fourth day had only had rice and salt left.

On the day of the mission, I carried a CZ .308 with a three-round magazine that had a 9X scope that was just as old as the twenty-year-old rifle. When I zeroed the scope a couple of days before the mission, it was dead on after three shots.

We walked into the target area and met up with one of the Karen snipers named Dah Oo (Monkey). Monkey had already killed ten enemy soldiers in this area over the last couple of months. He joined our team of six security, one overwatch/control, me, and my spotter. Monkey led the way with a bandaged toe sticking out from his sandals. The magazine on his SW .308 bolt action from the 1950s was held together with black electrical tape and his scope cover was an old tattered green bandanna. As he cleared the land mines he had placed along his egress and ingress routes, I could hear the bullets in his pocket rattling around. He had a girl friend and she was the one who helped him place the land mines here after his last mission. It took us five hours to go two miles due to all of the land mines.

When we got close, I put on my Ghillie suit and climbed up the hill. At the target area we found (remains) of an SPDC boot that had been blown apart and two SPDC land mines.

We were on an opposing hilltop (from the enemy). I got into position, standing on top of a six-foot-diameter tree that had been felled by a mortar round. I braced my rifle against a nearly vertical tree branch. My spotter looked though his 7X binos and started calling off targets. We saw two enemy soldiers returning from shopping at the local village. They had rifles slung across their backs and their hands were full of shopping bags as they walked between troop trenches.

My orders were to take out any targets of opportunity, to repay them for destroying the village the previous month. I aimed at the lead guy as he marched the steep path up the hillside. My crosshairs hovered between his head and shoulder blade, just above his AK-47 as the trigger broke and my weapon recoiled. When my sights settled, my target's head had exploded and he flopped to the mud. His blue T-shirt stood out in sharp contrast to the blood-stained soil.

Everyone (the enemy) paused momentarily as they tried to figure out what happened, then they dove into their trenches while I was chambering another round. The empty casing I ejected hit the tree branch and bounced back into the action as the bolt closed, causing a failure to eject and feed at the same time. I hurried to clear the malfunction as I watched the enemy through my scope scrambling from position to position. Feeling a little rushed, I fired at the head of an SPDC soldier as he peeked out of his machine gun position. In my haste, I underestimated the range, and the shot hit the dirt berm six inches below his head. He took off at a dead run for another position. He was carrying a Burmese rifle while wearing a white shirt and blue shorts and sandals. I got the range right and lead him by about thirty-six inches and fired.

I clearly saw the red mist blow out the far side as the bullet exited his torso along the lower rib cage. He flipped into the machine gun nest he was running for and two other soldiers went running out. He never got up. I had fired three shots in less than a minute and had shut down two targets.

Now the enemy was returning fire. They let loose with RPGs, M79s and two 81mm mortars, as well as every rifle they had. I was calling to my security and overwatch that "I still had targets," for they were yelling at me to get out of that position. My spotter had abandoned the binos and gone for his M16. That's when the KNLA "sniper," Mr. Monkey, started yelling for our security team to fire the M79. The snaps and cracks of passing bullets told me it was time to

The KNLA uses a wide variety of small arms. Some are captured from the Burmese thugs, and some are purchased on the black market.
(Photo courtesy of Jack Slade)

go and the more he yelled the more accurate the incoming fire became. The overwatch called for me to go. My spotter and I ran up the hill just as the bullets ripped through the leaves. We laid flat and watched holes appear in the wide blades of grass above us as if an invisible magic hole punch was working double time. We lay there until the small arms fire subsided and then we moved out.

Monkey re-armed his land mines as we withdrew. Within a few minutes, we heard from our radio intercept guys that there were two enemy dead.

Siege at a SPDC Bunker Complex

Another operation Jack Slade was involved in happened in late 2010, when the KNLA tried to drive the marauding SPDC troops out of the area by laying siege to their hilltop bunker complex.

Jack Slade: The SPDC, tatmadaw, junta, or whatever you wish to call them, had been trapped for several weeks by small arms fire, RPG fire, mortars, M-79, land mines, and ambushes from the KNLA and some of the newly allied DKBA units in the area. When I arrived on the scene, the SPDC were getting desperate for food and water. They tried to call over the radio to the other bases in the area for reenforcements and resupply, only to be told the other bases were under attack, as well. I spent most of the first morning on location creeping and crawling into position. I had my spotter with me as well as a Karen radio operator. My spotter was carrying a Colt CAR-15 A2 commando from the Vietnam era and the radio operator had a .30 cal M1 carbine from the Korean conflict. I was carrying a .308 Remington Model 700 with an adjustable variable power scope. We reached 800 yards and I could see enemy movement from trench to trench and bunker to bunker.

The Karen had positioned a Chinese .50 cal DShK machine gun on an opposing hilltop from the SPDC camp. We were in the valley below and the .50 cal would fire a string of lead at the SPDC and receive a volley of small arms fire back. As we got closer, the Karen radio operator kept telling us there were land mines near the stream at the base of the hill where the enemy camp was located. I reasoned that if I stepped in the buffalo tracks, I would be relatively safe from SPDC mines and went forward. However, the Karen radio operator refused to follow me. His eyes were filled with the fear of having his foot turned into a mess of shattered bones and burnt hamburger flesh. I don't blame him for being fearful. However, this meant that my spotter would have to move back and forth between my position and the radio operator to relay information about our position and hits.

I took aim on my first target holding a radio with a shiny antennae and binoculars on the top of the hill from a range of about 700 yards and hit him just as he was coming out of his command and control bunker. I believe the shot struck him where his neck joined his chest. At any rate, he was down and immobile after the shot.

At this time the Karens' .50 cal was nearly directly over head, and I was worried that if the KNLA gunners heard me shooting in the valley below them, they might shift fire and possibly hit me thinking I was part of an SPDC raiding party. So I sent my spotter back to the radio to tell Colonel Mya we were in the valley and not to shoot at us. Once my spotter left, four 81mm mortar shells dropped in the valley between me and the base of the enemy hill. They were close enough that I felt my ribs compress and vibrate. I could tell that they didn't know exactly where I was and were trying to scare me out of the valley. So I moved in closer. When I reached 600 yards away, I concealed myself in the tall grass and searched for targets while I waited for my spotter to return. I took note of bunkers, foot paths, trenches, vegetable gardens, structures, and planned for a long day of picking off the SPDC, one by one.

Then the KNLA fired some mortars and struck the command and control building, destroying the roof and the corpse I had left oozing blood at the doorway. The smoke coming from the area told us we had hit something of value. I ate a few crackers, drank some water from my Camelbak, and listened to the volleys of gun fire, RPG, and mortar fire echo all around me.

After a while I saw something reflecting light and took a look though my scope. Something inside a bunker was reflecting light. It could have been binos or a scope, so I dialed it in and fired. I fired at several more targets over the next two hours before the enemy moved to the other side of the hilltop to escape the accurate sniper fire and .50 cal machine gun rounds hitting their bunkers. After an hour of enemy inactivity, I decided to look for my spotter. Since there was little small arms fire coming from the hilltop and none of it was accurate, we walked silently back to the colonel's position to brief him on two kills and four possible hits.

That night was filled with flashes from claymores, M79 40mm grenades, and hand grenades as the SPDC soldiers tried to clear a path to the stream of any possible ambushes so they could make a run for fresh drinking water. Meanwhile, my spotter, two security element team members, and I moved into a position near the one I had used earlier in the day. I spent four hours picking off every light source I could see through my scope from approximately 550 yards. I could just make out cigarettes, the occasional candle or flashlight; they would come on until I fired at it, then it would go out. In the morning, our radio intercept guys heard that two more enemy had been shot dead, smoking kills.

At around 2:00 a.m., we moved back to the trench our command element was using and slept until 5:00 a.m. Then the noise of war awoke us.

I got out my rifle and scope and laid it on top of my pack and began look-ing for targets. I was surprised to see a patrol of five men moving from the stream back toward the tree line in the morning light. I estimated my range to the target to be 1,070 yards. There were two guys carrying five-gallon water jugs and three more acting as security. They were in a hurry.

I relayed what I saw to the command element and they shifted their .50 cal and RPG fire to that area. After eight rounds, the .50 cal malfunc-tioned. I fired at the patrol as quickly as I could to keep them pinned down and keep the water from reaching the thirsty troops up top. Then I saw one of the security element stop and turn. He aimed at the hilltop where the .50 was firing from and let loose a burst from his rifle. I placed my last mil dot on his head and fired. He was braced up against a slen-der tree, and I had enough time for my scope to settle before the bullet impacted his leg, hitting him in the upper thigh. He fell to the ground and crawled into a ditch. I saw one more and fired at him with the same hold over, but compensated for windage using "Kentucky windage." The round struck him in the shoulder and spun him like a top. On that mis-sion, I fired fourteen rounds and got five kills and two wounded.

* * *

These were just a few of the operations Jack Slade had taken part in, and as we walked through the jungle that day toward the KNLA village project, we came upon another face of war. Twenty-five or so civilians were hiding in makeshift huts along a small stream. The colonel translated as an old lady described how Burmese troops had burned her village and tortured her husband. Most of the children here had swollen cheeks, signs of malnutrition. The KNLA troops with us handed out bags of dried noodles from their backpacks, which the children started eating dry.

Colonel Mya stated the life of these internally displaced people (IDPs) is difficult. "They're always on the run from the Burmese army" and their "children often die from disease and malnutrition."

These were the people for whom the colonel, his troops, and Jack Slade were fighting.

The scorching heat beat down upon us as we left the IDPs; we slogged toward the new village project being cleared by the colonel's troops. "We will build her a new village," he said, speaking of the old lady, "but it will take time."

Colonel Mya later said to me, "The Karens are like David" as he quoted from the Bible. "As Goliath moved closer to attack, David ran out to meet him. It doesn't matter how much the Burmese regime try, they cannot defeat us. Today we stand on our feet and one day sooner or later they will have to respect our rights."

9

ISRAEL

Interview With Chuck Kramer

By Dale A. Dye

Although much has been said—pro and con—concerning the Israeli drive into Lebanon to evict PLO terrorists in the summer of 1982, the IDF has been reluctant to discuss lessons learned by their troops during the biller fighting in Beirut. What has been revealed about tactics employed in "Operation Peace for Galilee" has focused on the use of annor and combat aircraft.

Few students of the conflict in the Middle East realize the IDF—at the urging of an American émigré named Chuck Kramer—used the anti-PLO operation as a test and evaluation opportunity for a new cadre of highly trained snipers. The chain of events leading up to employment of snipers in Beirut and their record in combat is an intriguing story and an insight into sniper operations around the world.

SOF: How did Chuck Kramer become a sniper to begin with?

KRAMER: I was at West Point in the early 1950s and I was small-anns instructor there. I shot a lot and this was just a logical extension. After moving to Israel, I got involved personally with the commander of the Israeli border police. They were forming their anti-terrorist unit. At that time the border police were all Israeli combat soldiers. I was put in charge of their northern-area sniper group, which I recruited from veteran border policemen. I trained them with weapons available to the police, which were Austrian bolt-action Steyr sporters with their own scopes. Everything was bought commercially. North of the border right opposite Lebanon was Rosh Hanikra, Hanita, and other settlements where the terrorists came in close and fired light mortars, Katusha rockets, or what have you. For days we'd lay an ambush for these guys. They'd come up real close with a stolen Mercedes Benz and park the car, open the trunk, take out the mortar and shoot, then put it back in the trunk and leave. I worked as an independent *y'hiedah* [Hebrew for unit] attached to the border police headquarters, which was outside of Tel Aviv. I had then got ten friendly with the commander of the newly formed antiterrorist unit and he absorbed our unit by osmosis. The border police commander felt that there should be a massive sniper infusion into the antiterrorist unit. We just started building this unit by trial and error based on everybody's past mistakes and actual field experiences.

SOF: When was this?

KRAMER: This was in the late 1970s. We put on a demonstration for the Prime Minister and he asked, "How many rounds do you use up?" I said, "Well, we burn out around two hundred or three hundred rounds a day per man." He asked if that wasn't too expensive. I said, "Well, Mr. Prime Minister, how many rounds a day is your wife worth?"

SOF: What sort of training did you put the antiterrorist snipers through?

KRAMER: I made the targets as human as possible. I changed the standard firing targets to full-size, anatomically correct figures because no Syrian runs around with a big white square on his chest with numbers on it. I put clothes on these targets and polyurethane heads. I cut up a cabbage and poured catsup into it and put it back together. I said, "When you look through that scope, I want you to see a head blowing up." Everybody was satisfied. In retrospect, the time I spent in the antiterrorist unit and in the army was the only actual time I felt I was doing something positive for the defense of the country directly. It was an experience.

SOF: Now how did you switch from border security operations to the Army?

KRAMER: At the airport I had met "Rafoo" Eitan. I hadn't seen him in years. He said, "I heard some good things about you. Why not quit this kid stuff and come to the Army." So inside of a week he had become chief of staff and with three phone calls I was in the Army.

We reinitiated the Army sniper course, which was a total and absolute disaster. They were using outdated Czech bolt-action Mauser rifles made back in the 1930s. Their concept of sniping was completely antiquated. The Army *had no snipers* but on paper they had a lot.

We started from scratch with a semi-automatic sniper rifle modified from a Russian AK-47 manufactured in Israel, the M26 Circis. My own personal theories were that a sniper must maneuver with the smallest unit possible as the Russians do. Their sniper schools have never been closed since 1936. [In the IDF] there was no sniper doctrine per se. The Egyptian sniper school was very British-oriented. The Syrians had bought a lot of Steyrs—thousands of them—and were equipping them with the best optics and night vision devices they could get while the Israelis weren't doing much of anything.

So you were facing two basic problems: tactical employment of snipers—which you felt was wrong—and chronic lack of equipment.

The 1947 Independence War in Jerusalem where the Jordanians were shooting across the border at the Israelis was what they remembered. The crux of the matter actually sat in their officer training schools. There was no sniper indoctrination. They didn't ask the right questions. How do you employ them [snipers] as another tactical tool for the small—unit commander? In 1947, in Jerusalem, they were shooting from the rooftops with old bolt-action rifles after watching for a month. But there was nothing cut and dried and no tactical doctrine involved. So my initial plan was to convince the Army to buy enough of these M26 rifles, collect them and start a real good, serious sniper school in the desert. The sniper would leave with his own personal weapon zeroed and rebuilt for him and return to an ongoing training program in his organic/original unit.

SOF: You weren't interested in the Remington Model 700?

KRAMER: Not whatsoever. It's just another bolt-action civilian sporter configured for military application. It just didn't work well.

SOF: What was the training designed to produce for the IDF?

KRAMER: I built into this course everything I felt was necessary to get top-of-the-line, first round–kill snipers who could take targets at 0 to 1,000 meters. I had about fifty applicants for this course. We took the youngest guys who had the most time to serve. I asked all the right questions. "We want you guys to look through the scope and just see a target and watch his head being blown apart. Are you going to do this?" If there was any slight hesitation, I'd tell them to go home because I didn't have time to fool around. We shaved it down to fifteen guys because we only had fifteen rifles at that time.

SOF: What were the tactical changes that you wanted to make?

KRAMER: I would make a combat sniper who would maneuver with the smallest possible infantry unit—an eight-or nine-man squad—and indoctrinate the officer in the tactical use of snipers as another tool for his small unit.

SOF: What was the course like?

KRAMER: We were working by hook or by crook down by the Tzahal Officers Training Course Base No. J, living in tents out in the desert for thirteen weeks or so. We were independent, but officially attached to the infantry paratroop school. We built our own 1,000-meter range on a desert plateau. Anything I wanted, I did. I drew 2½ million rounds of ammunition and we started shooting. I told the students, "If you guys won't kill anybody I aim you at [with the] first round, then out." There was an *de corps* because they were getting good at this.

SOF: What weapons were you using?

KRAMER: We used old weapons. We used Jim Leatherwood's old ART scopes. I had targets especially made for us in desert brown, which were totally invisible to the naked eye. We changed the target configuration for the entire Army using realistic human-size targets in colors that the enemy is supposed to wear in the desert. At the end of the course, these guys would acquire in twenty seconds a target at unknown ranges and fire one round into it and kill it, any range, any place and not under formal range conditions. They were usually under a pile of rocks on the side of a mountain, shooting down. They also ran the close-combat course from 2 meters to 30 meters [to train for] instinctive fire with a sniper's rifle. They were set to kill all comers with our own equipment. They told us nobody runs a close-combat course with a 14-pound. sniper's rifle and nobody's got semi-automatic sniper rifles. They told us nobody shoots moving targets with an SLS [Starlight Scope] at 200 meters in the head. But we did all that. We were as trained as you could possibly get. We made an accounting and determined during the course each guy fired about twenty thousand rounds.

SOF: What sort of point-of-aim did you teach the snipers?

KRAMER: The target presentation ran this way: at 300 meters, head shots mandatory; at 300–500 meters you had chest-high targets only; 500-700 meters you had of a target; and at a kilometer, a full-size figure. We did it that way because in combat that would logically be how an enemy would present himself.

If the head was available, the point-of-aim was the head. The general rule held that the killing zone was pocket-to-pocket from the top of your breastbone down your sternum or stomach, sort of a big rectangle. I said, "I don't care where you hit them within that area, you're going to kill them." I felt we're not shooting these guys between the eyes unless that's the only available shot. You're looking to *kill the enemy* so you put one round right between his second and third buttons and he's dead.

SOF: What was the role of the sniper in a tactical situation as you explained it to the young lieutenants who visited the sniper course?

KRAMER: That he [the sniper] is your long-range personal artillery above and beyond the range of your personal weapons. In any situation where you needed a long-range, accurate, first-round kill to survive in a sort of unexpected tactical situation, you had in your possession a tool with which to do this. It's not like in the States where you could pick up air support or artillery fire from base. It was not available. You had to suffer with direct tank fire if it [fire support] was around at all.

SOF: Tell me about a tactical situation in the Middle East where a squad leader or a platoon leader would need his sniper.

KRAMER: Let's say you have an infantry platoon maneuvering in the southern Negev and they get pinned down in a tactically disadvantageous position by long-range heavy-weapons fire: a .50-caliber machine gun or equivalent. They're shooting at you from 1,000 meters and just holding you down until they can get accurate mortar fire. With a squad-level sniper, a lieutenant could send this guy out to a good position of concealment to discover the source of the incoming fire with whatever optical aids he had. If nothing else, he could use his 6X or 12X scope to spot and then just eliminate the source of the extended-range fire.

SOF: Go for the gunner?

KRAMER: Yeah, or shoot very close to him. The sniper could probably kill these guys with three rounds or they'd run away and then you'd just go along your merry way. If the enemy was employing tactical snipers in place of heavy weapons, our guy would assassinate them at a totally extended range just like the Indians. Kill the radio operators first, then quietly shoot all enemy soldiers in the head at ranges up to two and three times the effective range of their personal weapons. It would be like Custer's Last Stand with seventy guys laying out there dead, each killed by one bullet.

We were discussing snipers used *en masse* as a force or snipers used singly on the smallest tactical unit. A corps commander could just deploy a massive group of snipers attached to a larger unit to hold an area of miles. The snipers could just mark a path or killing zone of one kilometer in depth in which nothing moves for whatever tactical or strategic advantages you wanted.

SOF: What were your equipment recommendations for this new cadre of IDF snipers?

KRAMER: That takes some background. A combat sniper works at ranges far above the accurate, combat-effective range of the personal weapons of the infantry trooper. So he is working where their business stops. He is working out at a half-kilometer and farther with deadly accuracy day and night to get one-round kills under any conditions. That was the foundation of everything I did. All my input and data bears this out. All we were lacking to pull it off was the equipment.

The reticle I designed was an optical ART ranging device. I built the ellipse and the center of the vertical based on human head dimensions, which are 27 by 15 cm. That is the basic size of a human head. You would fit the target into an ellipse until the target's head filled it [the ellipse]. Combat sniping is far afield from antiterrorist stuff where you know the exact range. The equipment should be geared to help a sniper hit targets of unknown sizes and unknown shapes at unknown ranges day or night.

SOF: So you had a cadre of trained snipers and a modicum of adequate equipment. Tell me how Chuck Kramer got involved in the 1982 war.

KRAMER: My reserve unit was attached to one of the divisions of the northern area command. War started without the snipers. The opinion on sniping was—and is—still based on the old British adage that snipers use static positions. Anyway. I had gone north with a reconnaissance company. We'd gone up to the eastern part of Lebanon to Kfar Choonah east of Jezinne. The heavy fighting [in that area] was just about over in June of 1982.

The political pressure had just about stopped everything. We were interested in infiltrating terrorists who were running away from Beirut to get back to Syria.

We were about 90 klicks up into the Bekaa in some little mountain town when I got a telex from the chief of staff [Rafael Eitan] ordering me to Beirut immediately. I ended up in the Beirut University up on the mountain, which had been taken over by the paratroop infantry commander. I was assigned to a paratrooper unit above the airport. It was very loose. They said, "Okay, Chuck, go out there and see what you can do." No orders, nothing. Just, "Go see if this is worth your while." Things had quieted down at that point. There was just a lot of artillery firing night interdiction into the area north of the airport and lots of PLO Katushas falling all over the place. The area was liberally sprinkled by small-shots fire day and night.

SOF: What did you discover about sniper position and observation in Beirut? Did you work snipers in two-man teams or singles?

KRAMER: Two-man teams for protection. They [the snipers] get kind of tired. You can't stay at an SLS for more than a couple of minutes. One protects the other's back. Even though they all carried handguns, that wasn't much help when a guy is standing in back of you with a Kalatch [Kalashnikov]. I just put two guys together who were at ease with each other and worked well together and who wanted to be a team. No one was forced on anybody ever. If there were any mavericks nobody wanted, we got rid of them. There were no lone wolves and after you do this for awhile, you know how to play the game with your partner. These guys would gather intelligence data as trained observers. That was part of their job. Infantry guys who are sent out to observe will always miss something where an experienced sniper will pick this up.

SOF: And these observation techniques scanning and spotting, the business of looking for certain tell-tale signs were all covered in training?

KRAMER: Most of the guys in my unit were older and had been through five or six wars plus various campaigns. They knew what to look for. Even if a sniper doesn't shoot, he has his hand in the game, pulling in intelligence data. We did that in Beirut for about a week. I had sent these guys out and we weren't supposed to shoot—just watch. They poured in a wealth of data, which even the intelligence people had missed.

SOF: What happened with your snipers then?

KRAMER: At the end of July, I said, "You'd better get my unit up there [into Beirut] again." I was told there are no targets in Beirut. They [the IDF] are using direct-tank fire. I said, "I've got thirty fully equipped guys and we'd better get into this thing again." I told them not to worry, that my guys would find plenty of targets. This was at the time when the PLO were cornered in West Beirut and in the port. Everybody knew that Israeli defense forces were massing to drive them into the sea.

We got orders to precede immediately to Beirut to the division headquarters. They [the snipers] fired about five or ten rounds just for zero, loaded on an old bus, and proceeded up the road. We pulled up to headquarters and there was total confusion about what to do with us. They were laying the final plan for the assault on West Beirut out for regimental and battalion officers and they had amassed a lot of power. I had never seen so many tanks in my entire life. They had this twenty-block area rigged out bumper-to-bumper.

They wanted to go in there and just push them [the PLO] right into the ocean. We were to take twenty snipers. I split them up into three units. I had about ten guys with me. It was just a big mess: a lot of small-arms fire. We were east of the museum about two blocks. The first night we were there, we found ourselves in the wrong place and got in the middle of this huge artillery barrage. We were on the ground floor of a six-story, unfinished building. It was one of many sleepless nights. Welcome to Beirut.

SOF: Were you at this point prepared to use your snipers in the classic tactical role?

KRAMER: Right. I was assigned an area south of the hippodrome, which was right opposite the French ambassador's house and had a lot of PLO infiltration. We were to make this area too hot for them to go into. The idea was to push them as far west into Beirut as we possibly could while the artillery just shot down buildings. I had the largest sniper force of the entire Israeli Army at that point. The numbers at the end of the war indicated there were forty-seven snipers employed in the entire action involving more than one hundred thousand troops. I had over twenty [of the snipers] under one command. All the rest were dispersed throughout the entire force that was operating in Lebanon.

We got most of our [intelligence] data from either [Christian] Phalange or Lebanese Army units. The guys went out in pairs and looked every day in specific areas. The word had gotten out to the other side. Word [of the IDF sniper presence] spread like wildfire and these areas did quiet down immediately because they knew somebody was looking for them. They realized there was an organized group of Israeli snipers looking for them and this didn't sit too well.

We proceeded to stay there about a month until the multinational force had come in. We watched the Marines come into the harbor and then the French who went into the hippodrome to the ambassador's house. We saw the PLO open up on the French. There were French paratroopers here and there. We stayed until the general election and we assumed there would be a total uprising by the PLO after Gemayel was elected unanimously. We felt at that point the PLO would make a concerted effort to either break out or start an uprising in the country. We alerted all my forces to contain this thing in a small area in West Beirut. We stayed there until there was some sort of plan to evacuate the PLO and we were given orders to stop shooting. Small-unit commanders said, "Rifle fire is going on so kill as many as you can." I said, "What I don't know doesn't hurt me and if you want to squeeze off a couple for *auld lang syne,* just go ahead and do it."

SOF: How did you control the sniper teams?

KRAMER: During this whole operation I would be in contact by radio with these guys and a confirmed kill [witnessed by two men] was recorded as "one down." I'd record the range for my own records. Most of the good hits were at extreme long range and most of the kills were made at 600–800 meters. These guys [the PLO] were getting very wary by then. Nobody was walking around and the only guys who were there were PLO. There were no civilians in that area. Then they wised up and were running around in civilian clothes and had stashed their weapons. We spotted a guy on a motor scooter who seemed totally clean but who had weapons stashed in five or six apartments.

SOF: Were most of these kills at 600 or 800 meters people in buildings? Were you shooting framed targets or were you shooting open targets?

KRAMER: Both. It was a matter of targets of opportunity. Targets appeared at almost any range but there wasn't much close-range stuff.

These guys were too smart for that. They would stay very clear and they were using motor scooters a lot. I got two on a motor scooter by sheer luck. I was watching this big wide avenue and sure as hell I saw this big motor scooter coming toward me about a kilometer away with two guys on it. They're both carrying SKS carbines slung over their shoulders, two shopping baskets full of food, and these canvas carriers for the RPGs on front of the motor scooter. I lined them up and shot one off by sheer instinct. It was a classic shot. The round must have gone through the driver and then into the passenger. They went out of my sight picture and with that I lifted my head up and said, "Hey, score two for the old man." Then I looked down the road and I saw another guy getting up to run across the street. I squeezed off one round but I didn't hit him that bad and he fell down. I just squeezed off a second round. It was sheer luck: In a matter of minutes, two kills.

I documented all the things we were lacking at the end of this month period [in Beirut] and we left after the Marines had settled into where they were [a triangular perimeter around Beirut International Airport].

SOF: What new things about sniping did you learn from your experience with the Israelis in Beirut?

KRAMER: The Israeli Army had no experience fighting in big cities. All our sniper training and indoctrination paid off. I was proven totally right. Let's concentrate on specifics. All my theories were driven true in Beirut as far as sniper effectiveness in big cities goes. I wish I'd had two hundred [snipers] instead of twenty. If the equipment was better, I would have had more opportunity to perform better. I found to my surprise the PLO had very poor snipers using equipment almost six or seven years old in comparison with modern Soviet equipment. They worked alone with no sort of training. They were partisan. They just went out to shoot as many Israeli soldiers as they could. I found out—even as bad as they [PLO snipers] were— the reaction of the Israeli soldier to being fired on by a sniper was terror. It was terrifying to feel that you were under a sniper's scope. Units were stalled for hours while the commander was screaming for air support to take out a guy firing a plain Kalatch with no scope at 150 meters right into his position. I found the Syrians had good snipers. They had good equipment and good training.

SOF: After thirty days with twenty men in Beirut, what was the number that you put down?

KRAMER: Most of the firing was done in the first three days. The other twenty-six or so, days there were countless short cease-fires that were constantly violated by the PLO. That would cause the fighting to flare up again. In the first three days, we had sixteen [kills] verified, all at extended ranges, thirty-two probables, and countless war stories. This all occurred in about seventy hours in the first three days we were there. The closest shot was 150 meters—a total fluke with a sniper rifle—where one of the snipers lined up on this guy and found to his horror he didn't have a round in the chamber. He pulled the bolt back on his M14 and eased it in and the bolt didn't lock. That happens with an M14. He lined up, squeezed and *click*. The guy he was shooting at heard the click and ducked back inside a building. The sniper jacked the round out, cursing in Hebrew, chambered another, and waited. An Israeli air strike came over Beirut and these guys [the PLO] ran out like kids to watch the airplanes. That's when he shot him.

SOF: Based on your experience in Beirut, do you think the IDF learned anything about the employment of snipers?

KRAMER: They have learned that you *can* employ snipers. They're not as glamorous as a lot of other things, but they work. I don't see any born-again Christians in this saying, "Wow, let's make this a first-rate effort to incorporate one sniper per platoon and really train these guys because this is the way to go." It's nowhere near that level where I envisioned it would have been after they realized the fact that for the huge amount of troops employed there were just forty-seven snipers in the entire Peace for Galilee operation. It may be just slightly better now but I doubt it.

SOF: What about lessons learned regarding sniper equipment?

KRAMER: You've got to understand the prejudices sniper advocates were up against in the IDF to understand that situation. Major Commanders were from the old school of "follow me, automatic weapons fire, close and kill; snipers lay on roofs in static positions." One of the northern-area commanders had gone up for some antiterrorist missions in Lebanon. He was a non-sniper person at the beginning and he was a convert when he left the infantry division paratroop school [which he once commanded] where snipers were *the thing*. His name was Yossi Koller and he was a hard-nosed ex-paratrooper that eventually got the ball rolling.

We had tested every theory imaginable with every piece of equipment we had and made recommendations. When we saw the M26 was an impossible dream, I took 5,000 M14s out of stores, had my instructors shoot for a week, and took the eighty best rifles I could find. I had a purchase order for thirty modified M14s, which I helped design myself.

SOF: What were some of the modifications?

KRAMER: We took a stock M14 and rebuilt the plastic stock to the configuration of the M26 and extended the cheek rest to where your head was kept up and the scope was set off to the side. We hung a suppressor on the front, mounted a Harris bipod, and tuned the trigger action slightly. That was the best you could do.

SOF: By the way, did you have any left-handed shooters?

KRAMER: No, but the fact that the cheek rest was designed in such a way that a left-handed shooter could just lean over slightly and fire comfortably would have made them no problem. My exec was a lefty.

He fired an M26 all day long. He would just lean his head over and fire it from the physical right side of the weapon rather than the left side with no inconvenience whatsoever. The open sights were available if you had an optics malfunction.

SOF: What scopes were you using?

KRAMER: They were these 4X tubes made by some Swiss outfit and totally unacceptable. We had the configuration with a vertical post that totally blacked out a [long-range] target. The increment arrangements were so coarse that one click either way at 300 meters and you'd be off your target. These were 100-meter and 50-meter scopes that were bought in the late 1940s and reconditioned six or seven times. They were ready for the scrap heap. We had a couple of Leatherwood's ARTs and I had one on my own rifle, which I had built from scratch. We had given each man a personal SLS. These were the big bombs used in Vietnam, which were then almost twenty years old and totally unacceptable for sniping. I had epoxied the mount attachment to the SLS, which made for a beautiful side-mounting, but the flexibility in the mount was disastrous. Poor accuracy way out there.

I designed a concept for an efficient night vision device in which visible optics are replaced with an electronic image, so you're not concerned if the sun is shining or not. You have a heat-generated image of the target with highlights on the heart and head. That's all a sniper needs. The off-the-shelf equipment is available. The hardware is there, but nobody wants to make it. The market is small. Nobody really is interested and there are not enough guys around who have the experience to push this thing.

The efficient use of snipers in combat or antiterrorist operations is really an unknown factor. What is catalogued or written as doctrine is thirty or forty years old not only in the Israeli Army, but in the American Army, as well. I feel at this point there is no good equipment made anywhere in the world specifically for combat sniping. Modified military weapons won't get it anymore. The caliber is wrong and the optics are wrong. It's all made for deer and bear hunting. The night vision devices are for multiple-use heavy weapons. Nobody makes the gear for true, serious combat sniping.

SOF: What's needed?

KRAMER: A semi-automatic rifle just for sniping. Not a 7.62 but a .338 Winchester magnum, which is heavier [in cases] where you back off your range to 1½ kilometers. Olympic grade ammunition. You know the Marines custom-reload their [sniper] ammunition. They pull out the M113 bullets and replace them with more accurate, precision stuff. ART scopes from 6X to 18X where at a half-kilometer, you've got a big head sitting in there ready to be busted.

SOF: Why suppressors for combat sniping?

KRAMER: Simple. The suppressor does a lot of things for you. First, you're getting your recoil down to the level of a .22. You squeeze off that first round and what you see in that scope stays right where it is. If you miss, you can get that second round out quickly rather than have the recoil just obliterate your sight picture and force you to line up all over again. Your target is not going to stand there and let you kill him a second time.

SOF: So the suppressor is more a recoil reduction device?

KRAMER: There's more to it than that. A sniper has problems with muzzle blast. The suppressor eliminates this totally. You're also diffusing the sound. You won't get this supersonic crack diffused completely, but you can reduce it to a degree where when you're standing on the other end you're not too sure where the round is being fired from. Also a suppressor eliminates muzzle flash at night.

SOF: What sort of other equipment do you recommend for combat snipers?

KRAMER: Well, in Beirut flak jackets were mandatory but I don't think you can use them. They're hot. You can't get the rifle in there [in the shoulder pocket]. I don't say they shouldn't be used at all, but a sniper should have a modified flak jacket or an *ahfoad* [Hebrew for vest], in which he would have an open area for stock placement, all his pouches in the rear so his front is flat, and he can get down in prone position, a built-in shoulder holster and a carryall pouch in the rear for binoculars or what have you. It [the standard issue flak jacket] bothers the hell out of a guy lying in the prone position for an extended period of time. Once he gets edgy, he's concentrating on being edgy and he's missing a lot of stuff.

The weapon's got to be designed where his line of sight is natural without lifting his head up. Most American, British, and German rifles shoot from the heads-up position. The equipment must be designed to be comfortable over long periods of time. The optics, the eye pieces, everything should be designed to have optimum user comfort. We're using nineteenth-century technology in the twentieth century.

SOF: Give us some wisdom for urban sniping.

KRAMER: You need very good surveillance devices. Upward and downward shooting must be stressed because guys will not shoot low either way. It's very difficult to explain to these guys the physics involved. You've just got to show them. Nobody had the class is shot from being up on the tenth floor and shooting some guy at a nice shallow angle half a mile away. Most of the real targets presented themselves almost level, slightly above, or slightly below the shooter at extended ranges.

One of the things I learned about sniping in urban areas was that you tend to position snipers where they can observe and cover likely avenues of movement. We used good, real-time data from the Phalange or the Lebanese Army intelligence, which told us these people hit this area by day or night, or they live there, or they're operating out of there. Then you get as close as you want to get, put yourself in a good area for observation and just wait until they present themselves. There's not much new in all that.

I feel between 500 and 700 meters is classic sniping. I'd rather have the guys out that far than playing cat and mouse in a window or under a door. Get him way out there where he can't shoot. That has been my rule. I really leaned on the long-range unknown-distance shooting. I had guys shoot locks off doors at 100 meters. Why go through the door and kick it in? You have four guys that shoot in unison at that lock at 100 meters and it goes right through the door. The guy inside can't do a damn thing about it.

SOF: How about sniping vehicle crews?

KRAMER: We performed a lot of exercises with the armor—to their chagrin—where we combined snipers with antitank missile or rocket launcher crew. Most crews will maneuver with their commander's hatch open as a matter of comfort in the desert. We set it up where the tank commander was replaced [in the cupola] by a three-dimensional head-and-shoulders target with a tanker's helmet on. At 700 meters, a team of snipers would shoot out all the tank commanders. Immediately, the crews would close their hatches. For that short period of time, their observation was limited. At that point, the rocket launcher or missile team would put a killing round into each tank.

We developed a system in the Marincs for sniping armor crews in which we took the driver if he was available. We figured someone would have to dismount and get him out of there to get rolling again and we'd have another target.

That can work, but look at it this way: You've got guys in a close area and usually the tank commander is sitting right on top of them with the gunner off to the left and the loader off to the right. If this guy [the TC] gets killed, he is falling down into the turret and tangling up that whole area for ten seconds or so. The driver—for a few moments—has no communication with anybody because he hasn't heard that shot. The confusion factor alone will just about disable that tank where a rocket firing crew can just about expose themselves totally and get one good round out there and kill the tank.

SOF: Snipers could effectively cover a wire-guided missile crew then?

KRAMER: Oh, yeah. The crew has more than enough time because you're loaded and all set to go and you're watching those tanks. When the confusion starts you could just about stand up or kneel in a classic firing position [to bring the missile onto target]. We're not talking about extended ranges. We're talking about infantry-carried tank-killing equipment with time to squeeze that thing off. [The snipers are] working with three or four tanks, killing all the tank commanders at once, which makes it even more confusing because nobody's talking to anybody.

SOF: I've heard you want to employ snipers against choppers?

KRAMER: We had perfected a technique against the British missile-firing helicopters where you'd have a screen of snipers forward of your tank positions. They could acquire these helicopters at 800 to 1,000 meters as soon as they came up to sight the tanks. They usually hovered for ten seconds to acquire a sight picture for the AT missiles. The sniper's job was to present accurate, semi-automatic fire into the right-hand side of the canopy. Whoever was aiming the missile from that position in the cockpit would be totally occupied with staying alive and that pilot would get the hell out of there. If you fire ten rounds [into the Plexiglas], it's like fogging up that entire canopy.

Helicopters in general are easy to shoot down with sniper rifles. The whole thing is put together with a bunch of very complicated, very delicate linkages. You give an experienced sniper team a crash course on how a helicopter is put together and they can squeeze out twenty incendiary rounds that will make that thing fall like a rock. They're firing at all the sensitive points.

Snipers can also raise a lot of hell with fixed-wing aircraft. We once made a simulated assault on an air force base that had Phantoms on it. We were a kilometer away from twelve F-4s on the ramp. We simply showed the snipers where all the good stuff was in all those Phantoms and they shot for those areas. You're shooting at this monstrous target 30 meters long—you couldn't possibly miss. We told the base commander that snipers had destroyed every aircraft on his base. It was quite a lesson.

You can disable a couple of hundred million bucks worth of attack aircraft for about twenty bucks in sniper ammo.

Israeli Master Sniper

How the Israeli Defense Forces Became What They Are Today

By Lt. Col. Mikey Hartman (Ret.)

Lt. Col (Ret) Mikey Hartman was born in Memphis Tennessee, and grew up in Los Angeles, California. He dreamed of becoming a sniper in the Israeli Defense Forces (IDF) after he saw the movie *Raid on Entebbe* in high school.

Mikey moved to Israel at the age of eighteen, enlisted the Givati brigade (one of the five infantry units of the IDF), and never stopped amazing his fellow soldiers with his shooting ability. "The first word I learned in Hebrew was *Tzalaf* (sniper). I told my platoon commander I wanted to be a *Tzalaf*."

Mikey recalled his first time shooting with the Galil assault rifle. He fired a five-shot group (the IDF zeroes with five rounds and not three rounds like the U.S. military) of 1 cm, which back in 1988 was something extremely rare in the IDF. The next week, the IDF sent a general to watch him shoot and Mikey was sent straight to sniper school. "I was given very few gifts from God, and if you ask my wife, she will tell you even fewer than I think. But shooting and teaching people how to shoot is the gift God gave me," Mikey said.

When did you know you had the gift of teaching soldiers how to shoot?

"When I was about a year and a half in the army (men serve a mandatory three years while women serve two), we were getting ready for an ambush in southern Lebanon. I had just finished zeroing my then sniper rifle (M14 with a Nimrod 6X scope) on the firing line when a friend from my squad was having trouble zeroing his Galil. So I went over and told him what I thought he needed to change. My platoon commander saw this and said, 'Keep on going down the line and teach the rest of the guys.' After our tour in Lebanon, the battalion commander reached out to me and asked if I could try to develop the marksmanship training for our brigade. I trained all of the Givati units for about three years, and after we won six shooting competitions, the other units in the military started to complain that it was not fair that Mikey was only in one unit.

"When the commander of Givati brigade (then a colonel, now General Ret. Eiland) called me into his office and said he was with the general responsible for manpower of the IDF, he said to me that 'even though you are a master sergeant, we need you to go to officer school and build the shooting school for the Israeli military.'"

Historically, the Israeli army was not an accurate shooting army. In the 1982 war in Lebanon, called "Peace for Galilee," they fired approximately ten thousand rounds for every dead terrorist. LTC Mikey Hartman changed that. "In 1993 I became the head of the new shooting school for the entire Israeli military. I had two to three officers, five to six instructors, and that was about it. Where did I start? I sat down and started writing manuals that eventually became the doctrine by which all of the Israeli military trains. I would go to sleep and wake up with an idea of a new training scenario and write it down on a pad by the bed; the next day it became part of the training plan of the IDF."

Were Israel shooters that bad and how did you make them this good now?

"Our problem back then was we were subpar on the range but were a lot worse in battle situations. The stress of having someone shoot back at you is not easy to get used to. There was a huge difference in our hit rate in the battle situations. I had to close that gap.

"I started to study all past engagements with the enemy (those I was involved in and many others) to learn what I could from those encounters, i.e., range, time the enemy was in front of us, speed of moving target, how much of the enemy could be seen (shoulder, head, silhouette), what position we were in (prone, kneeling, standing, sitting, in between positions, behind obstacles), how many rounds we fired, day or night, and so on. I then designed shooting scenarios that best integrated all known studies and then built more shooting scenarios and [wrote] manuals that allowed us to get to the required drill in the right way (you must walk before you can run). We changed the entire boot camp firing drills; we threw out any firing that was irrelevant to battle situations that the IDF faces. I wrote over ten thousand different drills depending on what unit (tanks, engineer, intelligence, infantry, special forces, and so on) and what threat that unit faces. I also built a sharp shooting instructor course, which evolved from a three-week course to the nine weeks it is today. Finally, I quadrupled the number of marksmanship instructors in the IDF. In a matter of only six months, we began to see a huge improvement in both range and real life battle situations."

The IDF moved from the Galil to the M16 and now to the Tavor, all in a span of fifteen years, while you were at the helm; how did that work out?

"The Galil was a work horse, but much too heavy, and so are the Galil's magazines. We decided that the weight the soldier carries is critical for dealing with the enemy and staying alive, so we moved to the long M16A1 and very soon to the short M16. We didn't believe the advantages of the longer barrel were worth the length and weight difference. Try getting out of a BMP carrier or clearing a room in the territories with a long M16; not good, not good at all. After about two years, the short M16 began having serious jamming problems that actually cost the lives of some soldiers. After much begging and groveling on my part, I was able to persuade the IDF to purchase the M4 from the United States. At first Israel purchased only the upper receiver, but in the end they purchased the whole weapon. This was around thirteen years ago and it was a game changer. This reminds me of a nice story. My team and I have trained thousands of US military personnel before they deployed to Iraq/Afghanistan. They used to come to my base with their long M16 and I would always ask, 'Why do you guys still use this, why not move to the M4?' They would joke with me that the M4 is not good for ceremonies (too short). The M4 was the work horse for the IDF for about seven years; there are still thousands in use."

Other than the regular infantry weapons, were you involved in the introduction of any other weapon systems in the IDF?

"Yes, there are two main weapon systems in the combat divisions. One is the designated marksman [rifle] and the other is the Negev light machine gun.

"In 1997, we began to implement the designated marksman with a weapon called A2E3. In the IDF, we called it A3 for short. We used a Trijicon ACOG 4X for day and an Akila 4X for night. Over the years, the designated marksman became the most lethal position in the IDF infantry, responsible for more than 75–80 percent of all kill shots. Over the last decade, we have almost tripled the number of designated marksman in every infantry company. I spent many hours discussing this topic with the US military in the late '90s. It was such a learning experience for me to sit down and discuss the strategy on how an infantry company needs to be equipped with soldiers of our best (and unfortunately one of the only) friend, the USA.

"The Negev machine gun (Israeli made) was chosen as a light machine gun with a semi-automatic option for ambush scenarios to reduce the weight carried by the infantryman. Weight reduction is extremely important in the IDF and has been worked on extensively in the last twelve years. The 5.56 SS109 round has been proven to be effective at ranges that are relevant to the IDF doctrine."

What is the main difference between the US military and the IDF beside size and equipment?

"I honestly believe its discipline; in my eyes, the difference between the two militaries is huge in the discipline area. You guys are much more strict and go by the book, where in Israel we are the champions of improvisation. I'll tell you a quick story. Ten years ago I was in Camp Lejeune in North Carolina, and I spoke to a master sergeant about shooting technique. We discussed how you guys train soldiers to deal with stoppages in your weapon and how we do it in Israel. He said to me: We use the word SPORTS:

S—Smack/slap the magazine
P—Pull the receiver
O -Observe
R –Release bolt
T –Tap the forward assist assembly
S—Squeeze the trigger

"I looked at him and said, 'You know that that will not always work; the jam will not be fixed if you don't take out the magazine first in some instances.'

"He looked at me like I was crazy and said, 'I know.'

"I then returned the look and said, 'Then why do you teach it that way?'

"He responded, 'Because it's written that way in our doctrine. To change the doctrine,' he said, 'would take years of red tape and bureaucratic nonsense.'"

"I looked at him amazed and said, 'You know what the difference is between our two militaries? If I go to an IDF infantryman and say do this, A-B-C as it's written in our doctrine, he will say 'No, I have a better idea; let's try D-E-F.' That just about sums up the major difference in our two militaries.'"

In the IDF you use female marksmanship instructors; do the guys that are battle tested have a problem accepting instruction from women?

Initially yes; the military is a very chauvinistic group. Back before women were in combat, in the IDF it was very difficult for women to achieve the objective of teaching soldiers how to shoot. The initial response was always "how do you know, did you shoot anybody, were you in a battle," and so on. Over the years, I have learned that the women are better instructors than the men (in the IDF about 50 percent of all marksmanship instructors are women). They have more patience and better teaching skills. You must understand that the women go through extensive training before they become marksmanship instructors. Without the women instructors, the IDF would shoot less accurately.

Were they really called "Mikey's Angels"? There is also a rumor that they are all beautiful; how did that work?

When a beautiful young woman enters a firing range, the infantry guys shut up and do everything she says without questions; it just works better. You can teach professionalism; you can't teach beauty!

10

AROUND THE WORLD

SEAL Snipers Whack Three Pirates

By Harold C. Hutchison

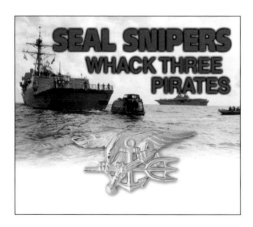

The guided-missile destroyer USS Bainbridge (DDG 96) tows the lifeboat from the Maersk Alabama to the amphibious assault ship USS Boxer (LHD 4), in background, to be processed for evidence after the successful rescue of Capt. Richard Phillips. Phillips was held captive by suspected Somali pirates in the lifeboat in the Indian Ocean for five days after a failed hijacking attempt off the Somali coast.
(U.S. Marine Corps photo by Lance Cpl. Megan E. Sindelar)

Information from Voice of America, the American Forces Press Service, and the Navy News Service was used to assemble this article.

Maersk Alabama **Attacked**

On April 8, the container ship *Maersk Alabama,* which was delivering five thousand tons of food aid to three African countries, came under attack from pirates as it neared its destination port of Mombasa, Kenya. The crew was able to retake the ship—but the pirates were able to take the captain of the ship, fifty-three-year-old Richard Phillips, hostage, retreating to a lifeboat from the vessel they had tried to seize.

"Captain Phillips's brave crew of civilian mariners fought back and took one of the pirates hostage and took their ship back," said Vice Admiral William Gortney, commander of the United States Fifth Fleet.

It was the first time since the nineteenth century that an American vessel had been seized by pirates, and the first time that one had been hijacked off the coast of East Africa. The piracy epidemic off Somalia had finally touched the United States. How had it happened? The *Maersk Alabama* was 300 miles off the coast when the pirates boarded.

Maersk Alabama *Capt. Richard Phillips, right, stands alongside Cmdr. Frank Castellano, commanding officer of USS* Bainbridge *(DDG 96) after being rescued by U.S Navy forces off the coast of Somalia. Pirates held Phillips hostage for four days.*

Ongoing, Changing Epidemic

Jim Wilson, Middle East correspondent for Fairplay International Shipping News, told VOA News in an interview before the hostage crisis was resolved, "The Indian Ocean is just truly a vast sea space. If you're 300 nautical miles away, then there's not a lot of help from a warship. Not even an air asset, like a helicopter, is [of] much value at that distance."

Indeed, all too often, Wilson added, "They actually take them by surprise. One of the best and first lines of defense is looking out the window. In this day of satellites and GPS and all kinds of technological gizmos, there's no real substitute for the human . . . eyeball. Look out the window."

Describing a Somali pirate attack for VOA, he added, "They come up fast and they'll do one of two things. Either they'll throw a grappling hook over (and) climb up. Or . . . they'll start shooting at you. Now, a lot of sailors will simply stop when they're confronted with pirates wielding AK-47s and RPGs."

Navy SEALs practice over the beach evolutions during a training exercise. SEALs are known for their ability to meet the mission objective in a clandestine way. Navy SEALs are maritime special operations forces that strike from the sea, air, and land. They operate in small numbers, infiltrating their objective areas by fixed-wing aircraft, helicopters, Navy surface ships, combatant craft, and submarines. SEALs have the ability to conduct a variety of high-risk missions, utilizing unconventional warfare, direct action, special reconnaissance, combat search and rescue, diversionary attacks and precision strikes.
(U.S. Navy photo by photographer's Mate 2nd Class Eric S. Logsdon)

U.S. Navy Response

The first American vessel to respond, the Arleigh Burke–class destroyer USS *Bainbridge* (DDG 96), was on the scene the next day, touching off a four-day standoff with the four pirates in the lifeboat, which had run out of gas.

USS *Bainbridge* was part of Combined Task Force 151, a U.S. Navy force sent to combat piracy off the Somali coast. Task Force 151's task is immense—it is responsible for more than 1.1 million

A Scan Eagle unmanned aerial vehicle (UAV) launches from the Navy SurfaceWarfare Center (NSWC) Dahlgren test range. USS Bainbridge used this type of UAV to monitor the situation during the five-day hostage standoff.
(U.S. Navy photo by John F. Williams)

square miles of ocean. With twelve to sixteen ships from a variety of navies, including the Royal Navy and the Royal Danish Navy, the task is daunting—and in the case of the *Maersk Alabama*, it meant help was 300 miles away when the pirates struck. In fact, Admiral Gortney stated in a U.S. Navy release prior to the *Maersk Alabama* standoff that over sixty ships would be needed to properly secure the area.

Bainbridge was only the first response. Additional naval vessels included the amphibious assault ship USS *Boxer* (LHD 4), which had not only helicopters that could carry Marines or Navy SEALs, but also some of the finest medical facilities afloat.

Failed Escape Attempt

The drama was heightened when Captain Phillips attempted to escape his captors on April 10 by jumping into the sea and swimming to the nearby *Bainbridge*. The pirates recaptured him and brought him back aboard the lifeboat.

The pirates began to move other seized vessels toward the area, as well, including a German container ship. However, the American vessels would arrive soon enough to set the stage for the dramatic rescue.

SEALs Take the Shots

As the evening of April 12 (local time) approached, tensions came to a head. The pirates were demanding a $2 million ransom and safe passage back to Somalia in return for the safe return of Phillips, now bound in the lifeboat, which was being towed by the *Bainbridge* to calmer waters due to worsening weather conditions.

"While working throughout the negotiation process tonight, the on-scene commander from the *Bainbridge* made the decision that

A team from the amphibious assault ship USS Boxer *(LHD 4) tows the lifeboat from the* Maersk Alabama *to* Boxer *to be processed for evidence after the successful rescue of Capt. Richard Phillips.*
(U.S. Navy photo by Mass Communication Specialist 2nd Class Jon Rasmussen)

the Captain's life was in immediate danger and the three pirates were killed," Gortney said. One other pirate, wounded when the crew had taken back the ship, had earlier surrendered to the U.S. Navy.

The instrument for saving Phillips: Navy SEALs on the fantail, who had earlier parachuted into the ocean and transferred to *Bainbridge*, took out the pirates with precision rifle shots from 25 meters when one pirate pointed an AK-47 at the captain—bringing about the end of the five-day-long standoff in a matter of seconds.

French Take Down Pirates

The day USS *Bainbridge* reached the *Maersk Alabama*, the French had acted to end a similar standoff that had reached its sixth day. A yacht had been seized on April 4, and five hostages had been taken by five pirates.

When negotiations broke down, the French stormed the seized yacht, saving four of the five hostages and killing two of the five pirates (the other three were captured). It was the third time that the French had used force to free hostages as opposed to paying any ransom.

For the Future

While the new Somali government welcomed the successful rescue, the pirates gained a new ally as a result of the actions: Islamic militants, including the al-Qaeda affiliated al-Shabab, hailed the four pirates as Somali national heroes. Some of the pirate gangs also claim that they are acting to prevent illegal fishing and dumping. Other pirates have now threatened to kill hostages. Indeed, pirate tactics had been ramping up prior to the *Maersk Alabama* incident. Insurgents also attacked the plane carrying Congressman Donald Payne (D-NJ) with a mortar.

"I understand now that from naval sources and also security risk management sources that the pirates are now approaching vessels . . . targeting the bridges . . . deliberately shooting out the windows in an attempt to intimidate the crew. And they've now taken to the habit of firing rocket propelled grenades...directly into the accommodation bloc (crew quarters). The idea there is to start a fire. If a fire is started on board a ship, it's exceptionally dangerous for the crew. So they have to stop the defense of the ship, which means deploying fire hoses . . . and put the fire out. When they do that, the pirates come on board," Fairplay analyst Jim Wilson told Voice of America.

But ultimately, the solution may not come from efforts at sea. "[T]he ultimate solution for piracy is on land. Piracy around the world stems from activity where there is lawlessness, lack of governance, economic instability; things of that nature," said Vice Admiral Gortney in a press conference held after the rescue.

Spanish Legion Snipers

By Richard Lucas

Most armies in the world today use snipers in some capacity and the Spanish Legion is certainly no exception. Recently I was able to spend some time with the Spanish Legion observing their training procedures, and noted that they seemed to deploy snipers more often than other military units I've seen. In fact, it appeared that at least one sniper team was somewhere in the field with us during nearly all the training exercises. During the weeks I spent with the Legion, I was able to appreciate their innovative use of the sniper teams and got to see some interesting sniper training exercises quite unlike what I was familiar with.

A superlative sniper weapon, the model 95 Barrett .50 cal. bolt-action rifle is accurate at over 1,000 meters and has enough power to stop a vehicle. In the field, camouflage netting is used to cover up the massive rifle.

To begin, the Spanish Legion units we are talking here are the 1st and 2nd *Tercios* (the Spanish Legion's equivalent to a regiment), which are stationed in the North African Spanish enclaves Ceuta and Melilla. Both African Tercios are mechanized light infantry units, which make up the core of Spain's Rapid Intervention Force. The regiment is equipped with 81mm mortars and .50 cal. machine guns in its support and logistics company, but apart from antitank and antiair missiles, that is the extent of the heavy weapons available inside the regiment. When deployed they count on other regiments to provide heavy weapons support.

In recent years, the two *tercios* have been involved in several United Nations and NATO peacekeeping deployments, including the Congo, Bosnia, Kosovo, Lebanon, and Macedonia. Spanish Legionnaires also made up the bulk of the Spanish contingents sent to fight in Iraq and Afghanistan. A lot of the training exercises I saw involved small, highly mobile combat groups and, again, most often with sniper support. The operational scenarios, such as hostage rescue

Spanish Legion sniper with a Barrett .50 cal. sniper rifle. Here he is carrying the cut down bullpup version, which is one foot shorter and ten pounds lighter than the full-size models.

and search and control situations, concentrated on urban warfare, the type of situation one would find in low intensity combat operations. In this type of operation, snipers, if they are properly deployed, can be extremely effective. Snipers operating at a distance with a good view of the operations area, and in radio contact with the ground unit, can lay down effective suppressive fire and take out selected targets, thus allowing ground troops to advance with the assurance that their movement is being followed and supported by the teams.

Riot Control

Most infantry or paratroop units, especially those that are involved in peacekeeping missions, will have one or more units trained and equipped for crowd control. Needless to say, they could also be deployed in domestic civil unrest situations, if need be, as support to police or National Guard-type units, which usually handle these situations. I have seen crowd-control training a number of times, but

Basic sniper–spotter team in full camouflage. The sniper has an Accuracy A1 AW chambered in 7.62 NATO. The spotter has a modified HK G36. Three-man teams are occasionally deployed.

here with the Spanish Legion, it was the first time I've seen a sniper team actively involved in the exercise.

This type of training usually consists of having the intervention groups disembark from the personnel carriers or whatever vehicles they are using. Dressed in full riot control gear—helmets, face masks, shields, and body armor—one group covers the other until they can position themselves and form a tightly knit skirmish line, which can advance behind the cover of the high impact plastic shields protecting them from rocks, bottles, or other projectiles. As they approach the crowd, they pick up the pace, stomping the ground, banging the truncheons on the shields as they chant in cadence. This can be fairly intimidating to the people they are approaching.

The confrontation line is backed by soldiers armed with rifles capable of firing rubber projectiles, gas, or smoke, followed by armored vehicles proceeding slowly behind. Upon closing with the rioters, there are usually several options depending on the situation. In training, the practice is usually to disperse them by driving a wedge through the center or pushing them into the open and rushing, truncheons in air, in attempt to sow panic in their ranks, making it easier to disperse the group.

So where do the snipers fit into this scenario? This was my question as I saw the two ghillie-suited soldiers disappear up a nearby embankment to set up an observation position covering the practice area. The captain in charge of the exercise filled me in as to what they were doing there.

"We set up a sniper team or teams to cover the area and to help in cordoning off the conflict zone. One of the main concerns we face in this type situation is to isolate the problem and prevent more people from joining in the demonstration. We have armored vehicles placed in strategic positions and the sniper teams cover main access routes as well.

"When you're faced with civil unrest, the primary objective is to try control and defuse the situation with a minimum of force. You

The Spanish Legion deploys snipers in a variety of combat situations. Here two sniper teams of the 1st company, 1st Tercio, based in the Spanish enclave of Melilla, North Africa, pose with the three weapons normally issued to the teams: Barrett .50 cal., Accuracy A1AW, and a modified HK G36.

have to remember that, if this situation were to be in our own country, in Melilla for example, the people over there could be friends or neighbors, so the troops have to remain as calm and disciplined as possible and try to avoid an escalation in the violence. But even in the case of domestic unrest, we may deploy sniper teams. As the teams are in radio contact both with the vehicles and the riot control squad, they can provide information as to the evolution of the situation and any dangerous element which may crop up. For example, from their vantage point, out of the smoke and chaos of the actual scene of the riot, they may be able to pick out the agitators, the more aggressive individuals. If the antiriot squad can coordinate a move to rush in and apprehend these individuals and hold them, it could help to defuse the situation.

"When you are faced with this sort of civil unrest during a peace-keeping mission abroad, like we saw in the Congo or in Kosovo, the situation is quite different and can be extremely dangerous for our men on the ground. In this case there is a very real danger of having the situation degenerate into a combat situation, which the troops in riot gear are not equipped to handle. Here the support groups would have to intervene in order to give the riot squad time to clear out of the area.

"For instance, if a vehicle carrying armed men were to arrive, one shot from the Barret .50 cal. using armor-piercing rounds could stop it in its tracks, forcing the occupants out into the open without excessive collateral damage. If shooters were to engage our men, the snipers' role would be to suppress the incoming fire and give the men time to get back into the personnel carriers and out of harm's way. So for us, in this type of exercise, the snipers are an essential element in the way of support and observation."

The next day, the first company of the Bandera was involved in a live-fire exercise in which the snipers played a more active role. On the rifle range, about 100 meters from the targets, a series of

A Spanish Legion sniper instructor demonstrates the use of the "sandbox" training installation. Firing a CO2 pellet gun, snipers train in target recognition, communications, and accuracy under various simulated weather conditions.

blinds, representing walls and rooms, was set up. In this exercise the combat team was to proceed across the range, using the blinds as cover as they fired on the targets at each opportunity. The sniper team was set up on a hillside about 500 meters from the range, again in communication with the squad leader. Their role was to eliminate specific targets, using cinder blocks that would pulverize when hit, so that the combat team could see the results and continue their progression.

The officer leading the operation explained to me that he was in radio contact with the sniper team and designated the targets as the exercise went on. "We often use the snipers as support for the intervention teams because they are mobile enough to infiltrate with the team and still have enough power to stop a vehicle if necessary or puncture through a wall. In a live-fire exercise, we want to get the men used to firing real rounds in a combat situation. The fact that the snipers are also firing over their heads is a bit unnerving, but it's important that they are aware of the presence of the sniper team and become familiar working with them so that they can identify the shots from behind (coming from the snipers) as friendly fire."

The sniper weapon used in the exercise, and actually all the others, was the Barrett .50 cal. Model 95 bolt-action rifle with detachable magazine equipped with a 16-power scope. The accuracy and the antimaterial and antivehicle capabilities make it an excellent support weapon.

The other rifle used by Spanish Legion snipers is the Accuracy A1AW (Arctic Warfare) chambered in the 7.62 NATO (308 cal.) cartridge, produced by the British company Accuracy International. The rifle is also equipped with a 16-power scope.

Legion sniper team spotters usually carry a modified HK G36 in the standard 5.56 caliber. The modifications include replacing the built-in 1.5X scope with a rail on which a long-range 16X scope is fixed. A cheekpiece is added to lift the eye level and nearly all are equipped with a laser-sighting device. These modifications give the rifle the ability to reach out to longer distances, up to 500 meters, with fair accuracy, while the laser device can be used when the target is too close for the scope. Also the presence of the laser dot coming in through a window, projecting on a wall, for instance, can have a psychological effect on the adversary. The fact that the G36 has a much higher rate of fire than the other two rifles makes it important when the team has to lay down more sustained fire power.

In-Regiment Training

Unlike their American Army and Marine Corps counterparts, the Spanish Army has no single central sniper training school. Selection and training are carried out by the respective regiments. I spoke to one of the Legion snipers, who ran through some of the basic procedures of the selection and training. "I was chosen for the sniper unit because I was good on the range. I had the highest scores in my platoon so I was offered a chance to try out for the snipers. I got through the selection process, which included both physical and technical tests, and started sniper training. We spent the first couple weeks or so just getting familiar with the weapons, first the Accuracy 7.62 and then the Barrett .50 cal. One of the gunsmiths would come in from time to time to make sure whichever weapon we were issued was properly adjusted to the individual sniper using it.

"After that we spent time in the classroom getting down the basic spotter skills, learning the techniques for the necessary adjustments for the shot: range, wind, temperature, and elevation, as well as target movement and deployment and tactics. In the field we worked on concealment, stealthy approach, camouflage, and navigation skills. Of course, there was a lot of live fire on the range."

Training

The Spanish Legion snipers are grouped together as part of the support company. They are then deployed out and attached to combat units for training and combat operations. The regiment has about

At the sniper training center at the 1st Tercio in Melilla, two snipers pose in front of the unit's banner of two wild boars. The wild boar is one of the mascot animals of the Spanish Legion.

twenty sniper teams, usually set up on the traditional spotter–sniper concept, although three- or four-man teams are occasionally used. When not out in the field, they work together in a separate building that houses a number of training installations specifically developed for the sniper–spotter teams.

Melilla and Ceuta are separated from the Spanish mainland by the Strait of Gibraltar, which makes access to the larger military bases and their training facilities somewhat difficult for the two "African" *tercios*. In reality, the enclaves are extremely small. Melilla is not even five square miles in size, which is about a fifth of the size of Manhattan. As this limits the scope and range of military activities, the *tercios* have had to innovate and have come up with some interesting training techniques. I observed this when I visited the sniper training installations.

As we entered, the sniper training building had written over the doorway "*Un Disparo, Un Muerto*" (one shot, one kill), the universal sniper maxim. Inside, the teams had set up various training devices. One, which is in common use for indoor weapon training, is where a combat scenario is projected on the wall and a rifle equipped with a laser firing device is fired at the targets as they come into view. An instructor observes and scores the shooter's accuracy. These scenarios usually work on the good guy–bad guy setup where the shooter has to pick out and eliminate the proper target in the shortest possible time. This is fairly easy in an urban hostage type situation, but in the forest scenario where they had to pick out the enemy by his camouflage and/or weapon, it became more difficult.

The most unusual training setup was referred to as *el cajon de arena*, Spanish for "the sandbox." Exactly as the name implies, sitting across the room was a large sandbox complete with toy soldiers, ships, tanks, and planes, something a kid would dream of having. One of the sniper instructors explained how they used it in their spotter–sniper training.

Accuracy A1AW; Caliber: 7.62mm NATO (.308Win); Operation: manually operated, bolt-action; Barrel: 610mm; Weight: 6 kg empty, without scope; Length: 1120mm; Feed Mechanism: five or eight rounds detachable box magazine.

"For the most part we use this for working on communication and observation skills. In the sandbox you can find about anything you might see on a battlefield today. Everything from an amphibious operation to an airborne drop, with an array of vehicles and combat situations and positions. We give the spotter a list of targets and he has to communicate the positions, range, and all other necessary information to the shooter, who uses a CO_2-powered pellet gun (Walther 1250 Dominator equipped with a precision scope) to hit it. We use code words to keep talk down to a minimum, keeping it short and precise.

"With the box we can test observation skills. They look over the scene for a few minutes. They go out and we change things around. Coming back, they have to pick out the changes and report them back. We can change the lighting to mimic any time of day or night, so they can practice with night optics and other specialized equipment. We can even fog it. It's also a good practice tool for reconnaissance. They have to sketch out a portion of the battlefield using standard military symbols; when they hand it in, we can evaluate the accuracy of their observations."

In position, a Spanish Legion sniper with an Accuracy International A1AW covers advancing units during training exercises of the 2nd Tercio in Ceuta, North Africa.

On the floor above was a computer rigged to a Barrett .50 cal. and to the shooter himself. Here, a small target was placed at the end of the room about 25 meters away. When the sniper fired at the target, again using a laser, the computer would not only show the point of impact, but record and graph out various data like heart rate, breathing, trigger pull, and barrel movement both before and after the shot. The head instructor told me that by using the computer, they could pick up on any problems in a person's shooting and correct them. Fine-tuning someone's shooting or correcting a specific problem on the range might take hours and hundreds of rounds of ammo, but with the computer and the correct program most problems could be easily solved.

Most of the Legion snipers I met had seen combat, a few in Iraq and others most recently in Afghanistan. After each mission, they come back with new ideas to refine and improve their training methods and performance. I read where an Army Ranger sniper said, "It's not like you can read a book and go do it. You have to do it over and over, and if you quit doing it for a while you can lose your skills. It's a perishable skill."

The Spanish Legion seems to have taken this philosophy to heart and to have put it into practice. As one instructor told me, "It all looks like fun and games, and sometimes it is, but it's one way to the stay fresh, developing new skills and improving on the ones they already have. The Legion is one of the first response units of the Spanish Army. All these exercises help keep the men sharp, ready to take on any mission on a moment's notice, and prepared to get the job done."

SECTION FOUR: GENERAL SNIPING

11

BASIC TRAINING

A Designated Marksman is an *SOF* Favorite

By Sgt. Hans Ruehr

Sgt. Hans Ruehr, USA, is armed with an M14 with a Leupold scope and Harris bipod.

Dear editor of *SOF,*

Hello, my name is SGT Hans Ruehr. I have been shown in your magazine twice–JAN 2007 and then AUG 2008. It's been a kick to see myself in your magazine!

However, my name has been spelled wrong both times! Someone put a "v" where there should have been a "u"! That photo was taken by a combat photographer early in 2006 during a combat operation in Mosul, Iraq. I was a designated marksman with 1st Platoon B Co., 4-23rd IN, 172nd Stryker Brigade. I was taken out of the fight

April 22 of that year by a sniper with a Dragunov. I was hit in the back of the head through my MICH (modular integrated communications helmet). It's been two years since that happened and I still suffer certain effects from getting shot in the back of the head. At least I got to keep my MICH! Right?

I just want to thank you for showing my photo in your magazine. My family and friends have gotten a kick from seeing me in there! You don't need to put this letter in the letters section or anything like that, just wanted to say thanks! I also attached a few more photos just in case you need to depict a good-looking SDM in a future issue! JUST SPELL MY NAME RIGHT, DAMMIT!!!!!

P.S. I love *SOF*!

Doing the Lord's Work

Sarge: My apologies for the incorrect spelling. We're running your letter next issue. Would like you to identify the various items you are carrying . . . like sun glasses, knife, etc. Also, on your M14 . . . what scope, bipod, Picatinney rail, etc. Also, the names of the individuals in the buddy shot . . . did anybody get the thug that dropped you? Are you back in the States now? Best to you and your buddies. You're doing the Lord's work. Robert K. Brown

Iraqi Police Apprehend his Assailant

Thank you for your prompt response sir! As for what I carried–I usually wore Oakley M frame "clear" for night and UVEX XC Polarized for day. My knife was a CRKT A.B.C. My M14 had a Leupold MK 4 10x40 M3 scope, Harris bipod, not sure what rail system it had, it was on there when I got it. At first it had a wooden stock, but eventually I switched to a composite stock to save some weight. My buddies in the photo are (from left to right)

SPC Shifflett, SSG Hernandez, SGT Strong, and me. After I got out of the hospital, I went back home on convalescent leave and I got a call from my old platoon leader. He told me the guy who shot me wasn't an "issue" anymore. He didn't go into specifics over the phone, but later on through various channels I heard that he had been caught by the Iraqi Police. Apparently he had killed a few IPs, as well. I am still in the Army, stationed at Fort Richardson, AK. I'm working for the Command Group of USARAK. Thank you sir. I love what I do and I love the USA!

McMillan Range Experience

By Lt. Col. Scott A. Blaney, USAR (Ret.)

RKB squeezes off rounds from the .308 McMillan Prodigy. RKB was duly impressed with accuracy and trigger pull and regretted he did not have time to send more rounds downrange. As you will note by the box, Black Hills ammo once again provided excellent results.

I followed RKB around the NRA annual convention held at the Phoenix Convention Center like a junior suck-up monk following the Dali Lama.

The highlight of the convention for the *SOF* crew was the invite by Kelly McMillan to fire his latest string of tactical rifles at the range facilities built for Maricopa County's Sheriff Office.

Robert K. Brown, his brother, Alan, Steve Schreiner, and I joined other members of the media. On the way to the range, Kelly gave a detailed history of his various operating companies that use his stock production and modern ultra accurate production technologies in McMillan's firearms offered to the military, police, and shooting public.

The McMillan Group operating companies now include McMillan Fiberglass Stocks; McMillan Machine Company, a contract manufacturer of machined parts for aerospace, firearms, and other applications where precision tolerances are required; McMillan Firearms

Manufacturing; McMillan Tactical Products; and McMillan Hunting Products.

We arrived at the range and the temperature was a balmy 94 degrees F in the shade (of which there was very little), a lip-chapping 14 percent humidity and no wind (used the Kestrel® 4500NV Pocket Weather Tracker [www.nkhome.com] to take the readings).

Under the watchful eye of McMillan's Director of Training Mark Spicer, author Blaney gets acquainted with the McMillan Tactical Hunter. Spicer is a retired British Army sergeant major who spent the bulk of his twenty-five years of service in counterterrorist operations. Spicer is a qualified (8541) U.S. Marine Corps Sniper and German Mountain (Alpine) Sniper and served as the U.S. government's expert witness in the Washington, D.C. Sniper trials.

Now a word of caution to the faint-hearted, his was no "wussy" bench-rest range. This was down on your belly, in your best prone position, load, lock, and blast away. We also had the option to fire standing (off-handed and supported), sitting, and kneeling. For the faint-of-heart, Kelly did provide mats, however, the hard as woodpecker lips (WPL) types like RKB got down in the dirt (tiny rocks and sand actually) and popped caps. Each person took a turn on each rifle and was able to fire as many rounds as he wanted. Most fired ten to twenty, except on the TAC-50, where some whimped out;

but some WPL-types, (yours truly) burned through twenty rounds of 50 BMG (Lake City) ammo.

The firing orders were set. The lineup consisted of six rifles, the first three from the new Custom Collection of hunting rifles.

1. Legacy in .270 Win, with a 24-inch barrel (BBL) and a 1x10 right-hand twist and topped with a Leupold 3.5x10x40 scope. This rifle was easy to handle and manipulate. The action was smooth and positive; the trigger broke cleanly with no creep. I shot two five-shot groups and the groups were a little more than 1.5"—not pretty, but that was the first shooting with a hunting rifle since last November. Remember, all these rifles were fired prone, off a sandbag—no benchrest comforts here—just the same as when you are hunting, shoot off your pack, bergstok, or "what-the-hell-ever" you have to rest your rifle on! I was impressed with the ergonomics, obvious ruggedness, and overall performance; I am sure it would be a sub-MOA shooter off the "wussy" benchrest.

2. Prodigy in 7.62x51 (.308 Win), with a 22-inch BBL with a 1x12 right-hand twist and topped with a Leupold 3.5x10x40 scope. This rifle felt exactly like the .270 Win Legacy and looked the same, it operated the same and the only difference was a bit less recoil. Again ten rounds in two five-shot groups, fired prone off a sandbag, and again two (approximately) 1.5-inch groups plus a flyer. My overall impression was the same as the .270 Win Legacy—that would be a good overall hunting rifle for most all game less than 700 pounds. Ballistic tables will show that there isn't a huge difference between the .270 Win, .308 Win, and the venerable .30-06 (from which the first two rounds were derived). You will find people that claim to have killed Godzilla with the .308 Win, which is irrelevant. The reason I like the .308 Win is that if you happen to be in a pinch, there will still be plenty of 7.62x51 surplus military ammo around to practice with

and in a pinch have a stash of a thousand rounds or so, for barter purposes of course!

3. Tactical Hunter in 7mm Remington Magnum, with a 24-inch fluted barrel with a 1x10 right-hand twist and topped with a Swarovski 4x16x50 scope and factory bipod. This was another smooth performer, same as the previous two Custom Collection rifles. However, this is the legendary "slap you in the cheek" 7mm Remington Magnum. I don't care what you say, this round has a mean report and recoil, but with performance that has earned the love and respect of tens of thousands of hunters, but not me. I know the ballistics—it shoots fast, flat, and hits with a ton of energy—I just can't get over the snarl (report). I did not shoot this rifle as well as the .270 Win and the .308 Win; probably due to my less-than-enthusiastic love of the 7mm Remington Magnum. (I, like Kelly's dad, prefer the .300 Win Mag.) My impression of the rifle was the same as the previous two Custom Collection rifles. I would have liked to try this rifle with the same compensator McMillan uses on their TAC-.338. However, if you're a lover of the 7mm Remington Magnum, this is the rifle to make the perfect match made in Arizona (some call it heaven, too).

The next three rifles were from the McMillan TAC Series.

1. TAC-.308A McMillan Tactical Rifle—featured a 20-inch heavy, match-grade, free-float barrel (with screw-on end cap thread protector), a Leupold 8.5x25x56 scope, the adjustable A3-5 stock in olive with factory bipod. Kelly designed this rifle for police/military use in the urban tactical environment. This rifle was designed to be fired prone, the butt hook and near vertical pistol grip working together to pull the rifle deep into my shoulder for a very firm hold. The bipod was exactly the correct height and easily adjustable for some of us more "portly" shooters. Ergonomics were well thought out and implemented. I fired ten rounds and again printed 1.5 MOA

with a flyer (same story—I was finally getting hot and I don't mean my shooting). This is a rifle any police department or sniper team would be well served to employ.

2. TAC-.338A Lapua (with detachable box magazine) in a magnum action. The TAC-338A Lapua is designed for long-range tactical scenarios. It has a heavy, match-grade 27-inch barrel, topped with a Leupold 8.5x25x56 scope. The TAC-.338 includes a muzzle brake and does that baby work miracles! The McMillan A-5 tactical stock features a spacer system, adjustable integral cheekpiece, and flush mount swivel cups. The new folding stock conversion is available as an option. Other options include a suppressor, bipod, and night vision rail system. A thread cap is provided that protects threads when a suppressor or brake is not in use.

    When I got to this rifle, I was expecting something that had recoil and report that was a cross between the 7mm Remington Magnum and a .375 Holland and Holland and was I ever surprised and impressed. This rifle is a joy to shoot. It has recoil that is about the same as a stock .30-06 pushing a heavy bullet. That is where the comparison ends. The first five shots were less than 1.25 MOA—probably dumb luck, because I was not particularly trying to pop that type of group. The action was extremely smooth and coupled with the far less than I expected in recoil, I decided to shoot the next five rounds as quickly as I could manage. I asked the range safety officer to time me for the five shots. After eighteen seconds, and five rounds, I had printed a group of 2.5 MOA, which impressed me and, more importantly, the range safety officer (former USMC recon sniper). With that performance, I decided to let that one stand and then moved on to the 800-pound gorilla.

3. TAC-50, The McMillan

    TAC-.50 caliber has been issued to armed forces around the world for ultra long-range situations, both antipersonnel and antimateriel. It is chambered in .50 BMG with a match-grade 29-inch barrel with muzzle brake and topped with a Leupold 8.5x25x56 scope. The McMillan Take Down .50 Caliber stock is popular for its space-saving compactness and features a spacer system and adjustable saddle-type cheekpiece. McMillan has documented evidence that a sniper using this model rifle set a world record successful tactical shot at nearly 1.5 miles. We did not shoot that distance, but we did knock the hell out of the backstop at 100 meters.

I had never shot a .50 BMG rifle before, only the M2 "Ma Deuce" for familiarization during basic combat training at Fort Polk in 1970, and that was impressive. Now enter Mr. "I can shoot any gun and be cool." I hunkered down with that beauty and cuddled her nice and close, took aim, and squeezed the trigger and, "Crap! The safety is on." That was the start of my humble pie eating contest with myself. I told the assembled laughing-ass on-lookers to be quiet as I had to concentrate. So, back I go to cuddling the Rubenesque beauty. This time I have a good hold, target lined up, and started the launch sequence. Then . . . *KA-DAMN-BOOOM!* About that time, I realized my habit of shooting with my mouth slightly open to help mitigate the muzzle blast was not the best practice with the TAC-50, as nearly immediately my mouth filled with dust, #9-size pebbles, and sand. My immediate reaction—gagging and spitting out the muzzle blast-driven debris—started the laughing onlookers to cackle again—humble pie eating sucks! In an attempt to redeem myself, I promptly fired another nineteen rounds until they all just gave up and mumbled something about it being too damn hot to stand there and watch Mr. "I can shoot any gun and be cool."

I am heartened that Americans can still make some of the finest shooting iron on the planet.

Robert K. Brown Goes to Sniper School

By Lt. Col. Robert K. Brown, USAR (Ret.)

Steve Langford served as RKB's coach and assisted others on the firing line. Langford presented a seminar on optics as well as the theory and practice of using mil dots. Brown failed to absorb the mil dot concept, so on the 500-yard line Langford, in frustration, finally said, "Dammit Brown, just hold three and a half dots above center mass and shoot." It worked.
(Photo courtesy of Robert K. Brown)

A Rainy Day in Sniper School Land

Why I do some things, I really don't know. For instance, here I am at the San Bernadino Sheriff's shooting range on a rainy Monday morning at 0800 hrs (I thought I had gotten rid of that silliness when I got out of the Army), getting briefed on how I was going to be shooting in the rain the rest of the week. Well, not really. The chief instructor decided early on that we would cram all the indoor seminars into the first rainy day and pray for sun the rest of the week.

I blame this foolishness on my old hunting buddy and Tactical Project Development guru for Bushnell, Steve Langford, who slyly suggested I might be interested in attending a week-long police sniper course. "Come on Brown, you might even learn something." I bit, but really didn't know what to expect. But what the hell. Any trigger time is good trigger time. And I just don't shoot much running unless I go to some sort of formal school.

The last rifle course I had taken was from Jeff Cooper prior to my jaunt to shoot the last of the species in Tanzania back in '97. (Now, gentle readers, that was 1997, not 1897). And, yes, I had been on the Fort Leonard Wood Hi Power team, competing in the Fifth Army matches back in 1970 after I toddled back from Nam. After firing on the wrong targets a number of times and being severely chastised by the team coach, I decided I'd better stick with the good old M1911 .45 cal. Furthermore, there was much less chance of getting muddy.

RKB with his Savage and suppressor used Black Hills Ammo 175-grain match ammo. Gang banger hat was purchased at a local store, which also had some less than tasty $1.97 a bottle Chardonnay.
(Photo courtesy of Steve Langford)

Langford offered to loan me a sniper rifle, which saved me trouble of hassling it through airline security. He had a number of them to choose from, as befits a man of his position and employment. I selected a Savage heavy barrel bolt gun topped off with a Millett Tactical scope with a snuffer. Though over the years I had tested a number of suppressors, I had never used one for any length of time and was interested in experiencing how the system performed under controlled conditions. (I had an early model Sionics suppressor, courtesy of the flamboyant Mitchell Livingston Werbell, III, in Nam to fuss with. I managed to shoot it off the end of my M16 in the first firefight I got into, which dampened my enthusiasm. I probably had not screwed it on tight enough and launched the damn thing 10 yards toward the VC. I later recovered it and noted that there was no blood, so it did as much good as my praying and spraying.) Langford noted that it was the same system that the Philippine National Police had recently ordered. A couple of days later, a snuffer showed up that reduced the recoil of the Savage significantly. Landford pointed out that the suppressor ". . .was a screw on, low maintenance system which incorporated a baffle. Basically, you screw it on and do what you have do." Langford, who graciously served as my coach, observed, "It shoots about 10 inches high with the Ops Inc suppressor, which means you're getting more distance out of the gun."

Frigulti critiques antiterrorist unit somewhere in the Mideast.

Finding Our Zeros

We started out spending a fair amount of time finding our zeros. I had mentioned to one of the instructors that I had a gut feeling that some of these young officers had not had a lot of trigger time. Zero day proved that to be the case. However, by the end of the course, everyone was shooting respectable groups.

According to Police Training Consultants (PTC) website, its ". . . primary mission and objective have not changed since its creation in 2000. PTC's objective is to provide training that will ensure and promote safety, survival, and success in the field. PTC has continued to grow and add new courses of instruction, providing training to both the law enforcement community and the military. In addition to providing tactical courses to law enforcement agencies throughout the United States, PTC provides tactical and firearms courses to units of the United States Marine Corps and the United States Army.

Langford, right, and Frigulti critiqued each shooter's target after each stage.
(Photo courtesy of Robert K. Brown)

"Recently, PTC returned to the Middle East for the second time in two years, providing training elite antiterrorism units in the Middle East. PTC remains committed to providing firearms, and tactical instruction that is fundamentally sound, innovative, and proven by experience. PTC will continue to provide training that exceeds individual and department's expectations."

Frigulti was blunt, to-the-point, and cut no slack when a shooter committed some sin. Shooter in orange watch cap is Brent Coon from Palm Springs PD: won shoot off. (Photo courtesy of Robert K. Brown)

A Bang Bang Career

Ronnie J. Frigulti, president of Police Training Consultants and primary instructor, started out his bang-bang career with the USMC as an infantry platoon and company commander in Vietnam, 1968–1969.

After getting out of the Corps, he signed up with the FBI, where he served on the Los Angeles FBI Swat Team for fifteen years as a sniper, sniper team leader, and assault team leader. He was certified by the FBI as a firearms, defensive tactics, and police instructor for twenty-two years. He was appointed as the principal firearms instructor for the Los Angeles FBI field office and served in that capacity for seventeen years until his retirement from the FBI.

Upon retirement, he has continued to provide training to law enforcement and military personnel. In the summer of 2000, he established Police Training Consultants (PTC), which is a firearms and tactical training/consulting company dedicated to officer safety.

Frigulti points out the damage RKB did from 500 yards. RKB commented, "This was a great confidence builder to say the least, especially using someone else's rifle."
(Photo courtesy of Steve Langford)

The course I attended was, according to the PTC website, "... designed for both the new and veteran Sniper/Observer who desires to improve their precision rifle marksmanship skills. Both fundamental rifle marksmanship and advanced engagement tactics and techniques are presented. This course emphasizes performance as the only true test of learned behavior for the Sniper/Observer. Students will be challenged to move and shoot in and about urban obstacles while also required to engage targets out to 500 yards (457.2 meters). Both known and unknown distant targets will be available along with a moving target. Exercises requiring movement, target location, identification, command fire, timed fire, multiple targets, and stress will be conducted. A low light and night fire exercise will be included. This intense course will provide a review of basic marksmanship skills as well as development of precision fire in an urban and long-range environment."

At the 500-yard line, the "usual suspects" pose for an informal class photo. By the end of the week course, all students were shooting respectable groups, though some had a minimum of trigger time prior to the class.
(Photo courtesy of Robert K. Brown)

So what did I learn? Well, it's a lot harder getting up and down than it was at the Fifth Army matches. But seriously, you never can have too much time behind the trigger. Frigulti said, "You need to get to the range once a week to maintain your skills—at least." A lot of the young officers taking the course simply did not have enough support, i.e., time off to go to the range or sufficient match ammo provided by their departments.

I must admit I was taken aback a bit when Frigulti *strongly* recommended cleaning one's rifle after every twenty rounds. It was interesting to note that of the fifteen shooters, only one had a semi-auto on an AR platform . . . which did not perform all that well.

Frigulti got blooded in the Corps in Nam as a platoon leader and company commander.

I was puzzled by the constant emphasis on ever reducing your groups. I asked assistant instructor Bruce Parks about this. He replied, "Very simple. We know through experience that if you shoot a MOA group under controlled conditions, that in a stress situation that will probably increase to two and a half MOA."

Parks also repeated the key word, "Consistency, consistency, consistency." That's what hurt me. I had difficulty resuming the same position after working the bolt. I did manage to wade through that impediment and somehow managed on the last day to fire a five-shot sub-MOA (barely) group at 500 yards, which opened eyes of some of the young officers, some of whom could not get on the paper.

The Twenty-First Century M1A

This classic rifle looks far into the future!

By Gary Paul Johnston

Few battle rifles can top the success story of the U.S. Rifle, Cal. 7.62mm M14. After taking over a decade to be developed and adopted, it was soon to be replaced by the M16 in 1963 at the startup of the Vietnam War. During the recent and ongoing conflicts in Southeast Asia, after a long list of complaints about M16/M4 failures, the M14 has been increasingly taken out of "mothballs" and refurbished and upgraded for use in the Global War on Terrorism. There, as with its ancestor, the U.S. Rifle, Cal. .30 M1 ("Garand"), the M14 is proving to be one of the best battle implements ever devised, this time in the form of a precision rifle.

In the late 1960s, however, a little known project was begun to produce a semi-automatic *only* version of the M14 rifle. With Directorate of Civilian Marksmanship (DCM) competition in mind, a secret program was developed at the Great Springfield Arsenal in Massachusetts. At about the same time as the semi-automatic XM21 (M14) sniper rifle was designed, a significantly different M14 receiver was also conceived from the ground up as a semi-automatic rifle, much the same as that of the M1 rifle. Although the then Bureau of Alcohol, Tobacco, and Firearms (BATF) approved the prototype "X" numbered design for civilian use as the M14M (modified), the Army rejected it, and the project was reportedly cancelled.

What happened to the original drawings of that semi-automatic receiver is unknown, but not long afterward, Elmer Balance of Devine, Texas, founded the commercial company of Springfield Armory, and began to produce the first commercial semi-automatic variant of the M14. It was called the M1A. A few years later, Balance sold the company and the Springfield name to the Reese family of Geneseo, Illinois, and the rest is history.

The basic M1A is the rifle that started it all. Many of these were used in the Middle East before the M14 could be put back into service.

Whether officially or coincidentally, the Springfield M1A could have played a part in the military's recent decision to reissue M14 rifles. This is because, in the wake of the failures of the M16 and M4 in the Middle East, some troops were allowed to take M1A rifles into battle there, especially in the role of sniper rifles. Whatever the case, the "re-adoption" of the M14 rifle has spurred the development of upgrades, which have brought this 60-year old rifle well into the twenty-first century. Springfield Armory's M1A is "riding shotgun" in this venture, as every improvement to the M14 also works on its semi-automatic sibling. Let's take a look.

Vltor Weapon systems

Springfield Armory offers over forty variations of its M1A rifle. While most are equipped with the standard 22-inch barrel in both light and sniper configurations, several use the popular 18-inch barrel in the

The first short barrel was called M1A-A1 was known as the "Bush" rifle. The latest version is the M1A Scout, seen here.

Scout Squad version and even more types come with the 16-inch "SOCOM-16" barrel. Vltor Weapon Systems has developed a tactical package for the M1A, which Springfield Armory now offers as an option on its 16-inch barrel SOCOM II M1A. Called the CAS-M1A Stock System, this package consists of a special alloy rail handguard, the top rail of which extends back to lock into a replacement for the rifle's charger guide. It will accept virtually any optical sight configuration including Aimpoint, Eotech, or conventional and forward-mounted scout scopes.

Vltor also offers a streamlined version of this rail system, which is compatible with any M1A or M14 rifle. Called the CAS-V 14, this rail system can be used with any Mil-Std M14 stock, or in combination with Vltor's modified M1A or M14 military polymer stock. It houses a full-length Mil-Spec 1913 top rail, two moveable side rails, a combination sling and bipod mount, separate pistol grip, and fully adjustable Vltor E-MOD Enhanced Butt stock. The CAS-V 14 system will accommodate a myriad of optics, lights, lasers, and so

As equipped with Vltor's entire CASV-14 Stock system, the SOCOM-16 uses Vltor's E-Mod butt stock and E-Pod two-piece bipod and a tan ERGO Grip. On the top rail is a Bushnell 10X42mm Tactical Scope in A.R.M.S. 30mm rings and ThrowLever mount. On the muzzle is Gemtech's Sandstorm titanium .30-caliber suppressor.

on. It comes in several colors, and the color of any synthetic stock can easily be changed with Duracoat's do-it-yourself kit.

The CAS-V 14's aircraft alloy handguard uses hardened steel components to anchor it to the rear of the barrel, with the rear extension of the top rail anchored to a special base that replaces the charger guide in the receiver. The operating rod guide is anchored to the handguard by a steel yoke and the ferrule contacts the front of the rigid stock as normal.

Vltor's E-POD is a unique, two-piece bipod that is right at home on either the SOCOM or Vltor handguard of Vltor's CAS-V 14. Consisting of two separate, adjustable legs that clamp rigidly onto any side rails, the E-POD's fulcrums ride above the rifle's bore, causing the weapon to hang in the bipod instead of balancing above it, as with nearly all other bipods. The CAS-V 14's bottom rail will also accept the Harris Bipod and the GripPod with its own light rail which, together with a Vltor Ring Mount, positions a SureFire G2 Tactical Light to be operated with the support thumb. Vltor stock components are used by a number of special military units and law enforcement units.

This totally rigid Sage EBR aircraft alloy stock is a perfect fit for the Springfield SOCOM-16. On its slider tube is the genuine SOCOM Butt Stock available from Lewis Machine & Tool and on the Sage top rail is the new HDMR 3-21 x 50mm Tactical Scope from Bushnell mounted in Millett 34mm heavy duty rings. This outstanding scope comes with the new Horus reticle, which is a cinch to use, and the top mounted Laser Devices laser does not interfere with vision. Here, a black GripPod and SureFire light are used along with the Gemtech Sandstorm titanium suppressor.

Sage International

Another highly specialized stock system for the M14 and M1A is the MK 14 Mod 0 from Sage International. The centerpiece of a family of such stocks is the MK 14's chassis, a totally CNC-machined aircraft alloy stock designed from the ground up as a mounting foundation for everything an operator would need or want to mount on the rifle. If this sounds like a tall order, read on.

Rather than any sort of a conventional top handguard, the Sage Stock comes with a machined alloy ventilated handguard that totally clears the barrel and becomes one with the chassis, using six heavy hex bolts. Atop this handguard is an integral Mil-Std 1913 rail running full length. Fastened to the bottom of the chassis via three similar hex bolts is a polymer wraparound hand rest.

With an almost endless choice of butt stocks and other bolt on features, the Sage M14 chassis is known as the MK 14 Mod 0, Mod 1 and Mod 2, or the Enhanced Battle Rifle (EBR). In fact, because of its total modular design, the system continues to evolve to fill different needs.

The precise fit of the action into the Sage stock automatically beds these two components. The barrel group is rigidly anchored to the stock using a heavy, machined steel operating slide guide that replaces the original guide, and which is anchored to the bottom of the stock by a large hex bolt. This bolt holding the barrel rigidly in place ensures repeatable sub-MOA accuracy, and assisting this is another bolt mounted through the top of the handguard at the front. With the rifle assembled into the stock, this screw is turned down to

The newest butt stock option for the SAGE Stock is the fully adjustable PRS2 precision rifle stock from Magpul. This version uses a special Sage mount and is extremely rigid.

just touch the top of the protruding barrel to dampen vibrations and control "stringing."

The latest improvement to the Sage MK 14 and EBR stocks is a greatly improved rail extension, which not only connects the top rail to the issue charger guide, but also elevates the mounting platform by .5 inches for better scope clearance. Locked in place with two robust adjustable cam levers, this new mount is a true problem solver; it needed to be for the optic we chose to test!

Bushnell's New HDMR Scope

Sent especially for this test was the new HDMR Sniper Scope from Bushnell Optics. Using a new Horus Reticle, this fabulous scope has a 34mm tube, ¼ click adjustments, and side focus and has a 3–21x50mm objective lens! No lightweight, this scope is not objectionable for what it offers, and that is super clear vision that far exceeds the range of the .308 cartridge. With it, we received a set of Millett 34mm rings. With the scope mounted on the new robust Sage extension rail, eye relief was a cinch! You'll be hearing a lot more about this and other new optics from Bushnell, and don't be surprised if some of it comes from the military!

In wide use throughout the U.S. military, the Sage Stock will accommodate all versions of the M14 and M1A, from full-length heavy barrel sniper rifles to the 16-inch barrel SOCOM-16. This short barrel will not only effectively engage targets out to 600 yards, but is "happy" when suppressed, with minimal effect on trajectory.

Smith Enterprises

In the wake of the increasing demand for its upgraded M14 rifles for the military, Smith Enterprises, of Arizona, has undertaken the production of state-of-the-art Mil-Spec M14 components, all built to print. These include complete forged/CNC-machined M14 and M1A receivers, bolt, match-grade gas cylinder group, match-grade 16-inch (SOCOM) barrel and much more. The latter is of particular interest, as the U.S. Army has moved toward a 16-inch match-grade barrel rebuild for many of its 7.62x51mm NATO precision rifles, including bolt-action and semi-automatic precision rifles. Where the M14 is concerned, Smith Enterprise's new MOA SOCOM-16 barrel provides match accuracy to an effective range of 600 yards and beyond.

Made from its inception to match specifications, Smith Enterprises' 16-inch M14 barrel looks identical to the Springfield SOCOM-16 barrel, but is held to much tighter tolerances, as is the gas cylinder group, which was part of the package I ordered. Refinement of these components over their issue M14 counterparts is noticeable. Using the proven four-groove, 1-in-10-inch (254mm) right-hand twist, this barrel is headspaced with a match chamber prior to being finished. This is necessary, because of the proprietary salt bath nitride type finish Ron Smith has trademarked as M80HT, which would make final chambering impossible, because of the 60 Rockwell "C" harness it renders to all surfaces. According to Smith, a longtime U.S. Government contractor, it will be next to impossible to wear out the bore of a barrel so treated. In addition to building M14 sniper rifles and sound suppressors for the U.S. Military, Smith Enterprises manufactures a line of Mil Spec M14/M1A and AR-15/M16 components. Smith Enterprises also manufactures a 7.72mm NATO titanium sound suppressor, which it has supplied to several military units for many years. This suppressor is not sold to the public, but another one is.

Gemtech's Sandstorm

Smith Enterprises' match-grade gas cylinder group for the 16-inch barrel is redesigned especially for their adapter with ⅝" x 24 threads. This short adapter will accept any .30 caliber or larger screw-on sound suppressor, and a natural is the titanium Sandstorm suppressor from Gemtech, of Boise, Idaho. Weighing just 13.3 oz., this 7.8" x 1.5" suppressor has a 32 Db reduction in sound, and has less effect on changing the point of impact than a heavier steel suppressor. I've

Here a U.S. Army Sniper uses a Sage MK 14 Mod 0 Stocked M14 rifle in Afghanistan. Sage stocks are in wide use in the Global War on Terrorism.

been testing the Sandstorm on several rifles, including my 16-inch Smith Enterprises SOCOM barrel, and it is a superb piece of gear!

Springfield Armory's M1A family can be had from the factory with a number of stocks, such as wood and synthetic, including conventional bedded McMillan fiberglass stocks in woodland camouflage and others, as well as with several heavy precision match barrels. No matter what you want in an M1A, Springfield can supply it.

The Sage M1 Garand Stock

However, as fine as the original style walnut stocks for the M1 Garand rifles are, the Sage M1 Garand stock offers something quite unique. Made with most of the features found on Sage's MK 14 and EBR stocks, the M1 version uses a similar barrel-holding fixture and can be had in three configurations, two with separate pistol grips. In addition, any of the same butt stock configurations can be had on the M1 stock, including one version that uses a more conventional looking stock. I recently received a sample M1 stock with the MK 14 Mod

Not leaving the M1 Garand rifle out of the mix, Sage offers three versions of its alloy stock for this rifle, also made by Springfield Armory. Seen on this M1 is the Sage MK14 Mod 0 fully adjustable butt stock and mounted on the top rail is an Aimpoint Comp M4 Red Dot Scope.

0 fully adjustable butt stock system. As with all other Sage stocks, ours came with an ERGO Grip from Falcon Industries.

While the Sage M1 stock uses the same CNC-machined aircraft alloy and an almost identical bolt-on handguard with integral top rail, it cannot use an extension on the rail, as this would interfere with the M1's top-loading *en bloc* clip. For optics, an extended-range Scout scope is the answer, and the stock's handguard holds it just as rigidly as on the MK 14 stock. This stock will also accept virtually all the same accessories as the MK 14 and EBR stocks. Not only can the Sage M1 stock increase accuracy of an issue-grade M1 rifle, but also for those who want the ultimate in a foolproof precision stock for a match grade .30-06 or .308 Garand, the Sage M1 Garand Stock is the answer.

Range Time

While the M1A and M1 rifles cannot be bore-sighted in the usual manner, sighting them in using the Ultimate Precision Shooting Rest from Carroll Targets, of Montrose, Colorado, makes it simple.

Simply sight on center paper at 25 or 50 yards, shoot a three-round group, and lock the rest with the scope reticle or open sights on your original point of aim (POA). Then adjust the windage and elevation knobs or center the reticle on the center of the group. Then move to a 100-yard target, shoot a three-round group to settle the reticle, and start shooting, adjusting the sights as needed.

Testing of the M1A 16-inch barrel rifle was conducted at 100 yards with commercial .308 Win. match-grade ammunition using 7.62x51mm NATO ball as a control round. Not surprising was that any match grade .308 ammunition averaged about one MOA, with several grouping well under that, while GI issue ball ammo would not break 2.5 inches. Accuracy from our 16-inch barrel M1A was close using both the Vltor and Sage stocks, but the heavier Sage stock produced smaller groups overall. The addition of the suppressor saw a change in point of impact of only about an inch vertically, but different ammunition brought slightly varying results.

One element that truly made it all happen was Bushnell's new HDMR Sniper Scope. Packed with optimum features, including a fast side focus, this scope features the new Horus H59 reticle, which for me is far easier to use than the earlier Horus reticle.

With our issue grade Springfield M1 rifle, we tested only commercial .30-06 ammunition before and after changing stocks, and changing to the Sage stock last. Overall, the commercial .30-06 proved superior to GI ball .30 M2 ammunition, especially when shooting commercial match grade ammo. In the Sage stock, overall improvement in accuracy was about one-third, or reducing a three-inch group to two inches.

Yes, the M14 rifle has redeemed itself as a top notch self-loading rifle. As for the M1A, it never left that position. As a friend once said casually, "Everybody should have one." The same could be said of the M1. To keep your Springfield rifles and the rest of your guns, *join the NRA! . . . Do it now!*

Specifications:	Springfield M1A	Springfield M1
Caliber	7.62x51mm NATO (.308 Win.)	.30-06
Muzzle velocity (16" barrel)	2560 fps	2800 fps
Operation	Short stroke gas piston	Long stroke gas piston
Type of Fire	Semi-automatic	Semi-automatic
Barrel Length	16"	24"
Rate of Twist	four-groove, 1-in-10" RHT	
Feed Device	5-, 10- and 20-round box magazine	eight-round *en block* clip
Safety	Positive safety in trigger guard	Same
Sights (front)	Protected post adj. for windage	Same
Sights (rear)	Protected aperture adj. for W/E	Same
Stock	Wood or synthetic	Wood
Finish	Mil-Spec	Same

Urban and Long-Range Sniper Course

By Lt. Col. Robert K. Brown, USAR (Ret.)

Most of you have undoubtedly heard me snivel that it seems the only way I get decent trigger time is when I take a class. Over the years, I've been instructed by some of the best, including (not necessarily in order of importance) Tim Lau, Louie Arbuck, Clint Smith, John Farnham, and Jeff Cooper, as well as Army instructors whose names are long forgotten.

So when I got a chance to once again attend a week's worth of bang-bang with a long gun honchoed by former Vietnam USMC Nam company commander Ron Frigulti, how could I say no? Ron made his bones serving on the Los Angles FBI SWAT Team for fifteen years, where he was a sniper, sniper team leader, and assault team leader. He also directed the LA FBI Police Training Unit, which

RKB engages target at an unknown distance from inside the Sheriff's Academy shoothouse. This Arrnalite has a 20-inch triple lapped ceramic coated 6 grove stainless steel match barrel, with 5/8x24 muzzle threads, and 1"10 rifling twist.

was responsible for the tactical and firearms training of agents as well as instructing local law enforcement and the military.

I had taken his police sniper course the year before last, but this promised to be a tad different, as it was going to focus on snipers operating in a urban environment. Once again, it was being held at the San Bernardino County Sheriff's Academy range, San Bernardino, CA—land of rotten fruits, stale nuts, and Nancy Pelosi.

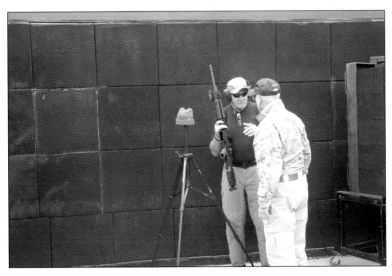

The Colonel and Ron Frigulti in a window of the live fire house at the San Bernardino County Sheriff's Training Facility. This window was used to place targets for live-fire sniper engagement into the house. In addition, sniper hides were used inside the house for live fire at unknown distance targets outside. RKB is holding an Armalite AR-10 Super SASS™ in .308 while Ron has a Remington 700 action trued with Sako extractor and large bolt, Hart SS 20-inch .308 barrel with a 1:10 twist, McMillan stock, solid bottom, pillar-bedded, barrel free floated, Jewel trigger, and a Leupold Mark IV 3.5 – 10 scope. Tactical rings and bases by Jim Gruning of Gruning Precision, Riverside, California. RKB wears a pair of Revision ballistic glasses and also a CAMPCO Uzi Protector watch.

Three Marine snipers, stationed at Camp Pendleton and friends of Steve Langford, were gracious enough to share some of their techniques with the class. The three had just come from competing in a civilian long-range match with their custom bolt guns. From left to right, GySgt. Shawn Hughes, Cpl. Dustin Payne, and GySgt. Tony Palzkill.

Getting an "assault rifle" was a problem, as the repressive California gun laws denied any out-of-state shooter, no matter how legitimate the purpose, from bringing one in. So, I contacted my old hunting buddy, Steve Langford. "Steve," I said, "I need your help—again." "What now, Brown?" I asked him, "What kind of assault rifle do you have that I can borrow?" We discussed what he had and I opted for an Armalite AR-10 Super SASS™ in .308 (no mouse gun for the old man). He had a Millett scope on it so I was ready to rumble.

The first day was spent getting zeroed and checking equipment. The course incorporated sniping techniques from hides built inside a house structure requiring live fire at targets outside the house and also firing from elevated positions outside of the house into the house structure at static and moving targets.

Urban tactical positions for live fire were utilized both inside and outside the house. The course included long-range target engagement at metal responding targets to both known and unknown ranges of 500 yards, requiring identification of target and range. A metal moving target was engaged from 100 to 200-plus. Range estimation, always a problem for me (yeah, yeah, I know . . . just guess how many football fields . . . which may work for Tim Tebow but it sure as hell doesn't work for me), was addressed with live-fire exercises based on the students' guesstimate.

I never like shooting an unfamiliar gun in a class, much less for more challenging projects. However, the Armalite AR-10 Super SASS™ served me well and I experienced no malfunctions using 175-grain Black Hills ammo.

I learned a lot, like how much I didn't know. Like cleaning my weapon every twenty rounds (I must admit, I cheated on this one). The impact of harmonics on accuracy . . . the pros and cons of bipods, etc. Langford even tried to teach me about mil dots, which I absorbed about as effectively as I have the way this damn computer

works. Finally, at the 500-yard range, he settled on, "Okay, RKB, hold it a little bit high and hair to the left." I managed to nail two out of four metal silhouette targets, which was better than a jab in the eye with a sharp stick, but not much.

In any case, Frigulti ran a tight ship with an able assist from Bruce Park, the department armorer, and retired Captain James Stalnaker, a legend within the Sheriff's Department.

The Armalite AR-10 Super SASS™ and I made a great team competing against the pros.

Marine Snipers Beat Brits

By Cpl. Jason Mills, 26th Marine Expeditionary Unit

Corporal Adam Harb, a sniper with the 26th Marine Expeditionary Unit, "talks shop" with a British Royal Marine Commando after participating in shooting competition. Two snipers from the 26th Marine Expeditionary Unit's sniper platoon participated in, and won, a sharpshooting competition against two British Royal Marine Commando snipers while aboard the USS San Antonio.
(Official Navy Photo by MC2 Joan E. Kretschmer)

Professional Contest

It's not easy to hit a small target bobbing along in the current as your helicopter shudders around you, 200 feet in the air. Add the pressure of international competition and things can get really difficult.

Two snipers from the 26th Marine Expeditionary Unit's sniper platoon recently participated in, and won, a sharpshooting competition against two British Royal Marine Commando snipers while aboard USS *San Antonio*. Both forces were in the Gulf of Aden supporting antipiracy operations when they took the opportunity for this good-natured professional contest.

Corporal Adam Harb, a sniper with the 26th Marine Expeditionary Unit, takes aim at a target on the ocean's surface more than 200 yards away.
(Official Navy Photo by MC2 Joan E. Kretschmer)

Not Sure Who Challenged Whom

It is unclear whose idea it was to have the competition—the U.S. Marines or the British Royal Marine Commandos, but both groups were more than willing to participate, said Gunnery Sgt. Jeffery Benkie, sniper platoon sergeant.

"There's two stories to it," he explained, "One is the Brits challenged us and then the other story is the Admiral challenged the Brits. So, I never was able to get a solid answer on that one."

Twenty-Five Rounds from the Chopper

The shooters fired on several different targets from a helicopter, each target floating at least 200 yards away and 200 feet below in the current. "The size of the target was probably about a five-gallon paint jug," Benkei said.

Each shooter had twenty-five rounds, Benkie continued, which they could fire from a weapon of their choosing, and the shooters could use any position or support they wanted to while inside the helicopter. Shooters earned two points if the round struck within a foot of the target, five points for a hit, and twenty points if that hit sank the target.

"It's a little different shooting out of a helo than it is shooting on the ground," explained Cpl. James Gosney, the Marine spotter during the competition, "because you've also got the rotor wash to take into consideration. Nobody really knows what effect that has. So you have to make your corrections based off the first shot."

Marines Win another Battle

Gosney and his teammate, Marine sniper Cpl. Adam J. Harb, came out on top 38 to 30. Benkie explained that after the shooters were done firing, they got a chance to meet their challengers when the British Royal Marines landed on the *San Antonio* for a brief meeting and congratulatory handshake.

Gosney said his favorite part of the event was after the actual shoot, when he had the opportunity to meet up and compare weapons and ideas with the British snipers.

Corporals Adam Harb and James Gosney, both snipers with the 26th Marine Expeditionary Unit, seek out a target on the ocean's surface more than 200 yards away.
(Official Navy Photo by MC2 Joan E. Kretschmer)

Exchanging Ideas

"We landed after the shoot and talked for about ten or fifteen minutes, and that was the best part of the shoot," said Gosney. "You know, in our work situation, we don't have a whole lot of chances to talk to people from other countries on our level, and especially other snipers."

"I think they learned a few things from us and we definitely learned a few things from them," Gosney said. "That was really worthwhile—[they're] really good guys," Benkie added. "It was very informal, but very educational at the same time."

12

WEAPONRY AND GEAR

Renaissance of Sniping

Innovative and Profitable Strategies for the Thinking Rifleman

by SSG Christopher Rance

Snipers are becoming an increasingly valued weapon in the mountains and deserts of Afghanistan. A definite renaissance of sniping is being seen along the front lines of war. The sniper was such an effective tool in Iraq. The snipers' tactical comeback in Afghanistan is facilitated by the mounting concerns over civilian casualties. Fear of collateral damage from coalition air strikes has made the sniper the military's most cost effective, discriminating fighting machine in the global war on terror.

As the combat force is drawn down and sequester cuts loom, the military will place additional dependence on force multipliers. Two of the most effective force multipliers currently in the war on terror are unmanned aerial vehicles (UAVs) or drones, and the modern

SGT Christopher Stevens shooting a LaRue Tactical OBR outfitted with a Bushnell HDMR riflescope.
Legion Firearms LF-15c with a Bushnell HDMR and PEQ-15.

military sniper. These force multipliers can be outfitted to emphasize any role necessary, from reconnaissance to combat and everything in between. Both are relatively inexpensive and fit well into the military's transformation to a smaller, more agile force.

Cost will always be a driving force in any military discussion. Statistics from past wars on the expenditure of ammunition to kill an enemy combatant are mind numbing. In World War II, the United States and its Allies expended 25,000 rounds of ammunition to kill a single enemy soldier. In the Korean War, the ammunition expenditure increased fourfold to one hundred thousand rounds per soldier. In the Vietnam War, the average number of ammunition expended per kill with the M16 rifle was fifty thousand.

In contrast, U.S. Army and Marine snipers in the Vietnam War expended 1.3 rounds of ammunition for each claimed and verified kill, at an average range of 600 yards. This figure shows the combat effectiveness of the sniper and the minimalization of risk to non combatants. In the current War on Terror, analysts have estimated U.S. forces have expended 250,000 rounds of ammunition to kill an insurgent. This figure also factors in ammunition used in training and combat. John Pike, director of the Washington military research group Global Security.org, has said that based on the GAO's figures, U.S. forces expended around six billion rounds of ammunition in training and combat between 2002–05.

Planning for sniper employment needs to be thorough and sound. The sniper needs to become a thinking rifleman. The sniper has to explore the uncertainties and inaccuracies of real-world shooting so he can make well-informed decisions about how to improve one's hit percentage. Bryan Litz, author and successful long-range shooter, has developed a modeling system that, if applied correctly, can change how snipers plan for future operations. The analysis method used in his book, *Accuracy and Precision for Long Range Shooting*, is called the weapon employment zone, or WEZ. Bryan states, "This is a systematic and comparative evaluation of small arms performance. The WEZ analysis is model based, statistical in nature, and it quantifies the hit percentage of a given shooting system on specified targets as a function of range."

The value of quantifying the hit percentage of a given weapon/ sniper/ammunition combination in a specific uncertainty environment is that the information collected can be used to quantify sniper effectiveness in war-gaming scenarios. Bryan Litz continues to explain the importance of quantifying hit percentage by comparison of several weapon systems under the same condition to ask important questions such as: How does a 10 percent BC increase affect hit percentage? How much is hit percentage improved by training a shooter to judge cross wind to within +/– 2 mph as opposed to +/– 4 mph? Are resources better spent on new rifles, or better ballistic software, or more training? These are all important questions to ask,

LaRue Tactical OBR and a Horus Kestrel meter in action.
SSG Rance, SSG Belford and SGT Stevens at the 2012 International Sniper Competition on Fort Benning, GA.

and by a thorough and careful WEZ analysis, the results can guide decisions as to which weapon systems and shooting skills are most effective at maximizing hit percentages in certain environment and confidence scenarios.

The decision makers need to be educated on how the sniper is employed, how he trains and what are his needs. They also need to understand how to calculate a meaningful hit percentage and decide where to focus resources and training. This role needs to be created and implemented in the U.S. Army sniper community.

The Army needs to establish a formal career progression for sniper-qualified personnel to become sniper employment officers through a warrant officer program. This position would create an incentive for sergeants and staff sergeants who have exhausted their sniper team leader and section leader time in a brigade combat team to become U.S. Army commissioned warrant officers. The move to adopt a warrant officer sniper military occupational specialty (MOS) would open the door for snipers to work in the field longer, receive better training, and improve the overall quality of the Army sniper.

The sniper employment officer would be able to advise and assist combat commanders on the employment, training, and equipment needs of snipers. This career field would be highly competitive and cost effective to the unit and the Army.

The U.S. Army needs to act. Snipers have a unique skill set that isn't being used to its full capacity. The demand for their skills and expertise continues to increase, but the incentive for seasoned snipers to stay is nonexistent. The framework and structure for a warrant officer sniper MOS would prove to be very successful and would continue the growth of the military's greatest force multiplier, the sniper.

The Noreen Bad News .338 ULR

By Gary Paul Johnston

A unique new semi-automatic .338 Lapua Magnum with MOA accuracy!

In 1983, during a search for a new military sniping round, a company called Research Armament Industries (RAI) began development of a new bolt-action rifle called the Model 300 and a brand new cartridge based on the .416 Rigby case necked down to .338 caliber. When RAI ran into difficulties in obtaining suitable brass casings and finances, it went to Lapua, of Finland, for assistance, and Lapua took over the project in 1984. The result was the .338 Lapua Magnum (8.6 x 70mm) cartridge introduced in 1989.

Having first tested the .338 Lapua Magnum in 1993, I was duly impressed with its power and accuracy. With a muzzle velocity of up to 3,000 fps and a muzzle energy of about 4,800 foot pounds, the .338 Lapua Magnum has an effective range of about 1,900 yards, all of the above depending on bullet weight and other factors.

The .338 Lapua Magnum rifle I tested back then was one of two prototype SAKO TRG rifles in the United States at the time. As with virtually all .338 Lapua Magnum rifles that followed, this one was a bolt-action repeater. Several unsuccessful attempts were made to produce a semi-automatic .338 Lapua Magnum. In 2007, a semi automatic .338 Lapua rifle, the MCR, was developed by Skip Patel of COBB Firearms. After Bushmaster Firearms and COBB Firearms were bought by the Freedom Group, Bushmaster obtained the patent to and produced a small quantity of the successful semi-automatic AR-based .338 Lapua MCR rifle, but decided (or were told) to shelve it. Now there is another.

The new Bad News .338 Lapua Magnum rifle will be anything but good news for terrorists. With this rifle's accurate range of 1,500 yards, our troops won't have to dial "0" for long distance.

Called the "Noreen Bad News .338 ULR (ultra long range)," the new rifle was designed and is manufactured by Peter Noreen of Only Long Range, Belgrade, Montana. An enlarged AR platform, like the Bushmaster MCR, Noreen's rifle uses a short stroke "tappet" piston instead of direct gas transfer, like the Bushmaster, but this piston operating system is different from anything now available.

The term "hybrid," as it relates to AR piston systems, is often mis-applied to anything different, but the Bad News rifle is a true hybrid gas system. As in a conventional Stoner AR, gas is tapped from the barrel into a long gas tube by which it is fed back to the front of the upper receiver. However, instead of being transferred into a gas cylinder/piston component (the AR carrier and bolt) to expand, the gas impinges into the front of a short piston housed in the upper receiver.

This "tappet" piston is in direct contact with an integral contact point atop the bolt carrier when the bolt group is in battery. When the piston is driven back roughly ½-inch, it imparts its energy to the carrier, which drives it to the rear, unlocking the bolt as it goes; the bolt travels to the rear via kinetic energy to extract and eject the empty case. It then returns under power of the recoil spring to feed and chamber a fresh round.

No Carrier Tilt

A situation common to any AR platform converted to piston operation (and they all amount to conversions) is carrier tilt from a piston slamming into the high point at the front of the carrier. This creates a strain on the components and will cause additional wear on the inside of the upper receiver. Noreen, however, has come up with a totally clever, out-of-the-box solution to the problem, at least in an AR-type rifle of this caliber.

Although the bolt and the front portion of the Bad New rifle's carrier are, by necessity, much larger than the same components in an AR-10 size rifle, the rear half of the carrier is still aircraft alloy and is the standard 1-inch diameter in order to use a standard AR-15

Johnston found the Bad News .338 Lapua Magnum rifle quite comfortable to shoot and extremely accurate!

recoil spring tube. While this sounds like a major "tilt" problem, Noreen uses a super tough, synthetic bushing housed in a recess at the back of the upper receiver and held in place when the receivers are closed together. The inside of this ring-like bushing fits perfectly with the diameter of the rear of the carrier and guides it straight into the recoil spring tube against the buffer. This bushing and the hybrid gas system of operation are the hallmarks of the Bad News rifle.

Made from one piece of 8620 steel, the 7-lug bolt is massive, as is its carrier. A cocking piece is retained by a spring detent to protrude out the right side for easy removal when field stripping is necessary. The 6061/T6 upper receiver houses a 26-inch Pac-Nor match barrel with a 1-in-10-inch right-hand twist. At the muzzle end is a very efficient muzzle brake. On top of the receiver is an integral MilSpec-1913 rail. The lower receiver houses a conventional AR-type fire control group, manual bolt hold-open device and magazine release. A match trigger is optional. Round or quad rail handguards are offered and a Magpul fully adjustable Precision Rifle Stock is standard. Both five- and ten-round single stack/single position feed steel magazines are offered. The Bad News .338 Lapua ULR weighs 13 pounds.

Our sample Bad News rifle came with a plain round handguard with a single 1913-type rail mounted at 6 o'clock, and on this we mounted first a standard Harris bipod with an A.R.M.S. Throw-Lever Mount, and later a short Harris Bipod with Harris's own cam lever mount in order to get the rifle low enough on the bench. The optic used was an SN-3 10X40mm Sniper Scope from U.S. Optics in the new A.R.M.S. #72 Magnum Harmonic mount with MK II micro adjustable locking levers.

This almost indestructible scope is one of U.S. Optics' top of the line optics with ¼ MOA adjustments, fast come-ups, illuminated reticle, and both rear focus and front parallax adjustments. A hold-over stadia is fast and easy to use and the glass is second to none. Made of aircraft quality alloy, this tough scope is hard-coat anodized and more abrasion resistant than most. I've used the SN-3 to test and evaluate a dozen rifles and it can't be beat.

Down Range

Ammunition used in the test was from Black Hills, Lapua, and Hornady. Although we were told that Hornady .338 Lapua Magnum ammunition would not work in the rifle, we decided to try it. In spite of the fact that cases stuck, we fired one three-shot group with Hornady .338 Lapua at 100 yards and it measured 0.96 inch. Black Hills .338 Lapua Magnum ammunition did not stick in the chamber and it averaged 1.84 inches at 100 yards. Since Lapua factory ammunition is the only brand recommended for the Bad News .338 ULR, that is what we concentrated on, and it averaged 1.10 inches at 100 yards with the smallest three-shot group measuring 0.84 inch.

Going to 400 yards, the Lapua .338 averaged 1.74 inches with the smallest three-shot group measuring 1.37 inches. In my experience, this is about as good as it gets with this round, and is the best accuracy I have ever seen from it in any rifle. What's more, I feel we were handicapped with a standard trigger and a 10X scope, although the U.S. Optics SN-3 is as about as clear as it gets. Peter Noreen tells me that the U.S. Army has made head shot-size groups at 1,000 yards. I believe it.

For a gas gun to outshoot a bolt action is quite impressive to say the least. This, together with its relative lightweight and even lighter recoil, makes the Bad News .338 Lapua Magnum ULR a rifle to contend with.

DRUTHERS

For me, the AR-15A2 type pistol grip is lacking and I would replace it with the ERGO Grip from Falcon Industries, the MOE Grip from Magpul or another. I also would have preferred the quad rail offered for this rifle so that I could use the EPOD Bipod from Vltor Weapon Systems. The last change would be a two-stage match trigger.

Darpa's XM3 Sniper Rifle

By Staff Sgt. Steve Reichert, USMC (Ret.)

A Marine Sniper fires the XM-3 off the bow of a ship while at sea.

For years, Marine snipers have been using their coveted bolt-action M40s (A1/2/3/5) to take out enemies all over the world. These rifles are all built by hand at the Marine Corps Precision Weapons Shop (PWS) in Quantico, Va. The Marines at PWS build some of the most accurate sniper rifles in the Department of Defense. Yet when 9/11 sent the Marines into Iraq and Afghanistan, the shortcomings of the new M40A3 were soon apparent. Marines on the front lines needed rifles that were shorter and lighter, had better fields of view, and were quiet. But the weapons procurement timelines in non-special operations forces are long, and what should take months typically takes years.

Enemy Snipers take a Toll

When Urgent Needs Statements (UNS) started coming from the sniper platoons on the front lines, one government agency was quick to respond—the Defense Advanced Research Project Agency, better known as DARPA. In 2003, DARPA became interested in helping the Marine snipers counter insurgent snipers who had been successful in spotting and killing Marine snipers in hide sites. When our nation's elite snipers are killed by enemy snipers, it tends to get the attention of all groups in DOD, DARPA being no exception.

DARPA looked at the problem from a technology standpoint. What could DARPA do to increase the Marine snipers' chances of survival while improving their lethality? One project they were working on was a sniper detection system called the Boomerang. In 2004, as this system was being field tested at Camp Lejeune against Marine snipers on Hathcock Range, the Marine Corps's representative to DARPA, Col. Otto Weigl, was on deck, watching the Marine snipers fire against the system. Col. Weigl noticed an older gentleman getting down in the dirt and talking to the Marines behind the rifles. The man asking the questions was Lt. Col. Norm Chandler, USMC (Ret.). He had been responsible for building Hathcock Range in the '90s, so he took an interest any time a new technology was tested there.

Lt. Col. Chandler had noticed a marine was having issues with his optics, and he asked what issues they were having in general with their currently issued M40A3s. That day on the range, Col. Weigl learned about the shortcomings of the Marines' sniper rifles and equipment.

Darpa Gets in Bed

I was the senior noncommissioned officer in charge (SNCOIC) of the 2nd Marine Division's Pre-Sniper course in 2004 when I got a call from Lt. Col. Chandler. I had known him for years and knew his company built some rock-solid rifles. Lt. Col. Chandler told me he

XM-3 on duty with MARSOC

had spoken extensively with Col. Weigl and that the colonel might be able to provide some technological assistance for the next set of units heading out the door.

I found it hard to believe that an O-6 would have taken such an interest, so I called Col. Weigl's office at the Pentagon. The colonel explained what DARPA could and couldn't do and asked to arrange a meeting with snipers and some key leadership within the division. That year DARPA held a small conference with Marine snipers to gather information on equipment and desired improvements. To develop a new program, DARPA decided to do an evaluation of off-the-shelf equipment that could be acquired, deployed, and evaluated.

The conference and evaluation led to many developments. "In mid-2005," Col. Weigl recounts, "DARPA provided a deploying MEU with spotting scopes, laser range finders, clip on night vision devices for weapons, carbine suppressors, and deployed two Mirage 1200 counter-sniper systems. DARPA also provided night vision and suppressors to the Marine Corps Warfighting Lab's first distributed operations platoon deployed in Afghanistan."

The initial equipment DARPA fielded was used in combat on a daily basis. The technical reports and after-action reports provided DARPA with the justification needed to start the M40XM program. Since DARPA had used retired Marine Lt. Col. Norm Chandler's Iron Brigade Armory (IBA - www.deathfromafar.com) to get the equipment to the Marines, they had a great working relationship. The folks at DARPA also understood what a battle-hardened rifle needed, so it was a natural fit for them to select IBA. In 2005, DARPA contracted IBA to build and test lightweight sniper rifles that incorporated the improvements the snipers desired in combat.

DARPA's mission was to develop a complete sniper system for both day and night operations. The system had to be lighter and smaller than the existing M40s, while having better accuracy, clip-on night vision that did not require a re-zero, better optics, and better stock, and it had to be suppressed.

Getting funding for the project was not an issue. According to Col. Weigl, "Funding for the prototypes XM1 through XM3, as well as the 56 full systems, was not a problem, since there was interest from the DARPA director Tony Tether, SOCOM General Brown, and

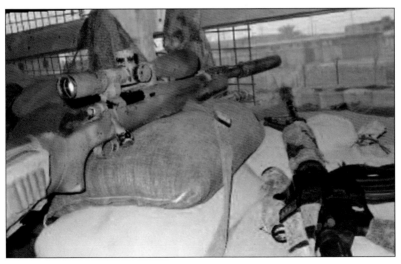

XM3 rifle in a sniper's hide somewhere in the theater of operations, next to a M4 carbine.

USMC General Mattis." The support and funding made it possible to expedite the development and fielding of the systems.

Developing the M40XM

IBA began development of the M40XM1 in early 2005. From the outset they wanted to develop a rifle that was lighter and shorter and that possessed a suppressor and night vision capability. Some of the issues with the Army's M24 and the Marines' M40A3s were long barrels, long actions (M24), Weaver rails (M24), heavy stocks (M40A3), and fixed power optics. IBA had to look at each issue on the M24 and M40A3 with a critical eye. DARPA wasn't interested in developing another M40A3 boat anchor.

IBA looked at all the parts in a standard Remington 700 action and began working to lighten and modify any factory parts to achieve better results. So what made the XM3s so different? Here are some of the main elements that set them apart:

- The receivers were clip slotted to accept the reverse-engineered titanium Picatinny rail (IBA design) to fit firmly.

- The receivers' internal threads were opened up to 1.070 inches to allow a perfectly true alignment with the bolt face and chamber/bore dimension. The chamber was cut to accept M118LR ammo.

- The titanium recoil lug was built with the 1.070-inch diameter opening for the larger-barrel threads and surface ground true.

- The stainless steel magazine box was hand fitted and welded to eliminate movement when assembled.

- The stocks were custom made for the project.

- The barreled actions were bedded in titanium Devcon and Marine Tex to allow for decades of hard use without losing torque or consistency.

During the development, IBA went through a total of five configurations before settling on the XM3's final configuration. The first and most obvious departure from a regular M40A3 was the stock. The Marines who used the M40A1 loved the sleek, low-profile stock. It didn't weigh much, it was easy to maneuver, and it fit most guys well. The main downside to the McMillan A1 stock was the low comb height—plus the fact that the fore end was not wide enough to accept the new in-line night vision mounts.

I called McMillan Brothers in Arizona and spoke to Mr. McMillan himself. I explained what the XM program was about and asked if he could take an A1 rear, raise the comb half an inch, and use an A3 fore end. He said it wouldn't be a problem, and they got on it. Within three weeks, I had the new A6 (as it was called at the time—it's now the A1-3) stock at my doorstep. Now the stock problem was solved!

The other major departures from the sniper rifles of the day were in barrel length and contour. The barrel had to be short enough to allow maneuverability yet long enough to deliver a 10-inch group at 1,000 yards. If the barrel was too heavy, maneuverability would decrease, yet if the barrel was too light it would only be able to shoot a few rounds before the groups started to shift due to barrel temperature. IBA tested a number of barrel lengths, ranging from 16 to 20 inches, and in different contours.

Each rifle with a different length was assigned an XM designator starting with XM1 through XM3. In each case, everything on the prototype rifles was kept the same except the barrel. During the final phases of testing it was found that the 18-inch barrels had no issues keeping up with their longer 20-inch brethren. The final barrel length was set at 18.5 inches, and the contour was a modified #7. The straight taper on the barrel was only 2 inches versus 4 inches and the overall diameter at the muzzle was .85 inches versus .980 inches. This helped reduce a lot of the rifle's weight while not negatively affecting accuracy or effective range. A number of the groups at 1,000 yards were <1 MOA.

Once the final rifle configuration had been settled on, the prototype XM3 was sent to the Naval Surface Warfare Center in Crane, Indiana, for testing, safety certification, and a comparison test. The tests conducted at Crane were very scientific. Every round was fired and recorded on a fully instrumented range. The XM3 was tested side by side with the Mk 13 Mod 5 and the Mk 11 Mod 0. The XM3 did extremely well during testing.

The time it took IBA to develop and field the XM3 rifle was light-years ahead of typical government programs. By the time the first XM3 rifle had been shipped out the door, only twelve months had passed since DARPA had contacted them.

XM3 rifle packed in a Hardigg iM3200 Storm Case. Also visible are some of the accessories included with the system, including the Nightforce NXS 3.5-15X50 MD scope and the Sure-fire FA762SS supressor. A cased AN/PVS-26 night vision system is also visible in the case.

The XM3 Rolls Out

You would think with such support, the Marines would be chomping at the bit to get these new weapons systems in hand. Despite the interest from Gen. Mattis, some Marine program managers at Marine Corps Systems Command (SYSCOM) said there were no requirements for a new sniper rifle and made DARPA jump through numerous hoops. The Marines tested and evaluated the XM3 at Quantico. SYSCOM required an official safety certification from the Navy's Surface Warfare Center, and once the bureaucratic pushback from SYSCOM, PWS, and unit armory and supply officers was overcome, the systems were sent to the units.

In 2006, the Marine Corps started to take delivery of the XM3 sniper weapon system. The system included:

- Rifle—XM3

- Hardigg iM3200 Storm Case

- Suppressor—Surefire FA762SS with soft case

- Day Scope—Nightforce NXS 3.5–15X50 MD with ZS

- AN/PVS-22 UNS night vision unit with soft case

- Harris Bipod—BRMS with PodLoc

- Eagle cheekpiece

- Turner Saddlery AWS sling

- Dewey cleaning rod and bore guide

- Seekonk torque wrench @ 68 in lb.

- Kleinendorst bolt disassembly tool

- Allen 5/32 T-wrench

- SK T30 T-Wrench

- Kobalt ½" adapter
- Three bore brushes

A number of the first units to receive the XM3 were West Coast infantry battalions. Col. Weigl had been working with GySgt. Ken Sutherby, one of the Corps's top snipers, to ensure the rifles made a smooth transition into the fleet. GySgt. Sutherby was instrumental in ensuring that the sniper platoons who received the XM3s knew how they operated and what they could and could not do.

This was the first time the Marines had seen in-line night vision devices that did not require them to be zeroed to a specific rifle. It was also the first time the Marines had had variable power scopes and most importantly the first time they were able to shoot their rifles suppressed.

Col. Weigl and Norm Chandler, Jr. were on hand at Camp Pendleton when the first shipment of rifles was delivered to I-MEF. As with all IBA rifles, the XM3s were test fired and zeroed before leaving the shop. When the Marines cracked open the cases and went to zero the rifles, they were pleasantly surprised that all the rifles were within a ½ MOA of their point of aim. The Marines were able to hit their targets all the way out to 1,000 yards with ease.

But that night was when the Marines' jaws really dropped. After the sun went down, the Marines tossed on the PVS-26 Universal Night Sights. Using their prior data, more than three-fourths of the Marines had first-round hits at 900 yards, fully suppressed!

The Marines loved the fact that the rifle was more compact and lighter and had more capabilities than their existing M40A3s. The rifle was mostly able to keep up accuracy-wise with the longer-barreled M40A3s. Even though the barrel on the XM3 was a full six inches shorter, muzzle velocity was only reduced by 100 fps on average. Did this make the rifle less accurate? No. It just meant that, depending on the environment, the rounds sometimes went trans-sonic prior to reaching the 1,000-yard mark.

XM-3 Goes to War

Shortly after I-MEF took receipt of the XM3s, the first units in II-MEF took receipt of theirs. By mid 2006, there were dozens of XM3s in Iraq quietly killing insurgents. One of the first reports back described a team of three insurgents emplacing an IED about 400 yards away from the team. It was just after midnight when the team shot the first insurgent. The other two had no idea where the shot came from and starting running directly toward the team. The sniper took out the second guy, while the third guy kept run-

Snipers and spotters on a shooting line train for the first time using the XM3 sniper rifle. Snipers found that they were getting rifles that were getting first-round hits on targets 900 yards away.

ning toward the team. The third guy was dropped about 200 yards from the team's position.

Reports like this were coming back from theater monthly. I thought the Marines themselves could best sum up their thoughts on the XM3:

> I last deployed in 2008 and carried the XM3 on every operation. The rifle shot great and the size was perfect for the insert platforms.
>
> —GySgt. USMC (MARSOC)

> I think the move to the XM-3 was great—beautiful gun, big step forward. Never had any problem with function (or malfunction)—awesome rifle, love it.
>
> —MGySgt. USMC (2nd Force Recon)

The length of the XM3 helped with employment issues as well. In an urban environment where you can get some standoff from your loophole, size is not such an issue. In the mountains of Afghanistan and in the small mud buildings of our AO, we were often wedged into our positions. A shorter weapon was welcomed.

—Sgt. USMC (Sniper Instructor)

Why we haven't adopted this model of rifle I have no idea. I love this rifle. If I could afford it, I would buy one myself.

—SSgt. USMC (Sniper Instructor)

The greatest advantage to this rifle compared to the M40A3 was the decreased weight, suppressor, inline night sight rail over the barrel, and variable power scope. That would not be significant nowadays, but back then, it was a force multiplier.

—CWO2 USMC (Marine gunner and
former Sniper School SNCOIC)

The length of the rifle was its real advantage. To be able to T-bone it across your pack and not have it sticking as far out as our 40s was advantageous.

—SSgt. USMC (Sniper School Instructor)

Everyone in our platoon loved that gun. Nobody ever got to take it out with them because Gunny was so possessive of it. He called it "his gun."

—Sgt. USMC (Sniper School Instructor)

At the time, none of our 40s were set up with suppressors. That, in my opinion, is what gave it the advantage over the M40s.

—GySgt. USMC (Sniper School Instructor)

I personally shot over 1,000 rounds through the XM3 and had no issues. With M118 I held sub minute of angle groups back to the 1,000-yard line, suppressed and non-suppressed. This was the norm and not the anomaly.

—CWO2 USMC (Marine gunner and
former Sniper School SNCOIC)

These benefits are hard to explain to a commander that only leaves the wire under the protection of an armored vehicle. We tried one night, but our battalion commander could not see how any military unit could justify spending that much money on a weapon system. We explained to him that a lot of our heavy weapons cost a lot more. This was to no avail. In the end, I left him with a playing card that I had shot the spade out of from 100 yards using the XM3.

—Sgt. USMC (Sniper Instructor)

All around, the XM3 was the solution for a primarily urban fight and the required long-range shooting when the opportunity presented itself. The suppressed capabilities allow snipers to remain uncompromised and take more than one shot from a concealed position.

—GySgt. USMC (Sniper School Instructor)

XM3 Issues

One of the few pitfalls of the XM3 program was that it wasn't official. Therefore, no structured training or maintenance plans were in place or ever implemented. This meant that if one of the rifles went down for whatever reason, its life was over. Although the XM3 could have easily been maintained by the 2112s, SYSCOM wouldn't officially let them work on the rifles, which reduced their productive lifespan in the hands of the Marine snipers. I know of a few 2112s who would do the right thing and fix any sniper rifle a marine is using, regardless of what paperwork is or isn't on file at SYSCOM—but this only goes so far when the official policy gets in the way.

One of the most respected Marine snipers holding an XM3 rifle. Based on the Remington 700, as was the M40 sniper rifle that was in widespread use by the Marine Corps at the start of the War on Terror, it features improvements over baseline M40 sniper systems, including a Surefire suppressor and a McMillan A1-3 stock that combined the best features of the stocks used on the M40A1 and M40A3 sniper rifles.

Another issue from a shooter's point of view was the optic selected for use. The optic was the same one used by the SEALs on their Mk 13s, but it was adjusted in MOA. When the XM3s were fielded, the Marines just started switching over to the SSDS with mil adjustments. The veteran snipers knew MOA adjustments, but not all really understood them since their Unertl 10Xs were single-revolution BDCs.

For a new sniper coming out of sniper school who knew mils, it was now necessary to also learn MOA. The scope also lacked the reticle in the first focal plane. This meant the only way to do correct mil readings or moving-target leads was to power the scope to its max. Some of the units used the Nightforces and did great with them, while others never could figure out MOAs and instead put on either old Unertl 10Xs or the new SSDSs.

Most of the XM3s became theater assets, meaning they were left in the combat zones and were transferred from unit to unit. This meant that an incoming Marine sniper was issued the XM3 upon entry into his area of operations. The Marine probably had no prior training and was relying on data someone else had gathered and that he had not confirmed for himself.

Seven Years Later

In doing the research for this article, I found that most of the XM3s have been sent to the Marine Corps Logistics Base (MCLB) in Albany, Georgia. Had there been a program of record, the 2112s could have worked on the rifles and kept them in service. As it stands right now the 48 XM3s in Albany are slated for destruction in July and August of 2012. Four rifles remain in the fleet.

The Marines could do a few things with the XM3s to ensure they continue their service in some way. First, they could wisely spend some end-of-year funds to get the rifles refurbished and upgraded by the manufacturer. This would add fifty-two more sniper rifles to the inventory, bringing the Marine Corps's total number of sniper rifles to near 1,200. The total cost to have all fifty-two rifles re-barreled, re-bolted, re-bedded, and upgraded to detachable box magazines would be just over $100,000.

Second, the rifles could be sent back to the manufacturer, where any worn parts that would cause safety issues would be replaced, and the rifles would then be sold in the manner that Remington is sell-ing the Army's old M24s to raise money for the Wounded Warrior Project. The third option would be to have every rifle chopped up and dropped in the dumpster.

To see these fine combat sniper rifles destroyed would be a shame. I hope the Marine Corps does the right thing and puts these rifles to use, either in killing insurgents or in raising money for our wounded warriors. I feel personally attached to these rifles, as I was involved heavily in their development. I'd still have my own XM3 today if I hadn't sold it to a Marine sniper instructor in Hawaii during a time of financial hardship. I hated to see it go but was glad it went to a fellow shooter. To this day, even with over six thousand documented rounds, he's still getting 1/2–1/3 MOA!

Spanish Sniper Puts Gear to the Test

By Jorge Tierno Rey

Assault rifles shooting station. Free shooting time with the HK 416 (5.56mm 14.49-in bbl) with RDS Aimpoint Micro T-1.

The fourth Aimpoint Live Fire Days took place in August 2010 in Sweden. Sponsored by Aimpoint AB, the Swedish company that manufactures electronic rcd-dot sights, this event was also supported by U.S. and European manufacturers of firearms, eye and ear protection, target systems, range robots, tactical clothes, medical items, ammo, flashlights, infrared (IR) lasers, scopes, and mounts.

Present were thirty-eight attendees from thirteen countries, all members of elite military or law enforcement units (mostly SWAT or SOF). We were divided into groups to carry out the different activities. Despite our different mother languages, we were able to share experiences together in English.

Big bore weapons shooting points. Shooting the HMG FN .50 M2HB-QCB with RDS Aimpoint MPS3.

We were lodged in the barracks at the facility, which is located by the sea at the fabulous Ravlunda Range in the southeast region of Sweden, away from urban areas. The firing ranges are set up to accommodate small arms fire out to distances up to 1,000 meters.

Monday, August 30

I arrived with the rest of the Spanish contingent at the Copenhagen Airport in Denmark, where an Aimpoint crew member was waiting to take us for the two-hour car ride to Ravlunda. Once we were assembled, the president of Aimpoint met us at the canteen for a welcome presentation.

After lunch, a small but focused manufacturers' exhibition was set up so that we could talk with each of the representatives about their products. One of the notable products was Saab Aerotech's PROLIFIC computer software, which calculates danger areas for all weapons and ammo involved in an exercise, and integrates the information in map overlays and safety orders.

MG shooting station. Shooting the FN Minimi 7.62mm with RDS Aimpoint Micro T-1 and 3XMag.

We then attended a safety briefing, which included a presentation on trauma wound treatment and some training with a few new medical products. The lecturer was a Swedish flight surgeon who provides medical training to units around Europe. We were given a medical kit to use in case of a real emergency during live fire exercises.

Tuesday, August 31

The morning started with a presentation of the different target systems we would be shooting. Canaxa Target Systems presented their full line of targets that included turning, running, and pop-up targets. Their radio-controlled pop-up targets are a good solution for any unit on a tight budget.

Saab Training Systems presented their advanced targets. Their multipurpose Stationary Infantry Target (BT 18SIT) can be radio-controlled by their Range Control System software, allowing one to design a course of fire with many targets and to record the performance of each shooter.

Sniper shooting station. Shooting the awesome .338 Timberwolf, manufactured by PGW Defence Technologies.

Later in the morning, the senior sales director and the president of Aimpoint presented their most advanced fire control system (FCS12). This sight integrates in one piece of equipment a red-dot scope, laser-range finder, ballistic computer, and inclinometer. This sight automatically adjusts the point of aim according to the weapon, ammunition (50 ballistic algorithms are available), range, and terrain angle. It is easy to manipulate—just aim, range with the laser, and aim again with the adjusted red dot.

Then it was trigger time! We had the opportunity to try all Aimpoint sights and accessories, including the Micro T-1, CompM4s, 3X Magnifier module, and Concealed Engagement Unit (CEU), which is used for indirect fire around obstacles. We started the exercise by zeroing from 25 meters, noting that the mechanical offset keeps the point of impact (POI) about 1 inch below the point of aim (POA) at close range. After zeroing the optics, we were to be given the opportunity to try out a wide variety of individual weapons. I positioned myself next to the FN SCAR-L CQC (10-inch barrel), so when the line went "hot," I could get that rifle! As soon as I shot the SCAR, I fell in love with it. There would be time to shoot each weapon later on the event. The day finished with what they called a Swedish Pentathlon, a group of team-building games.

Wednesday, September 1

It was time to shoot the M4 carbines. We carried out some exercises to check zero and learned that shooting with both eyes open is easy. To prove this, we fired a course with the front covers of the Aimpoint optics completely closed. Yes, you can do it! We also fired with the red dot in various positions of the sight to prove that there was no need to center the dot before firing—Aimpoint sights are parallax free. Then free fire time arrived. We could shoot any weapon as much as we wished. As you can imagine, a lot of rounds went down range! The toys included Heckler & Koch G36 KV (12.5-inch bbl.), HK416 (both 10.4-inch bbl. and 14.5-inch bbl.) and HK417 (20-inch bbl.). From Sig Sauer we had the SG 553 and SG 551. From FN the P90, F2000, SCAR-H (7.62mm. 16-inch bbl.) and SCAR-L CQC (5.56mm. 10-inch bbl.) Nammo provided all the ammunition we fired during the event.

Aimpoint shooters zero their weapons and optics at 25 meters.

Afternoon was time for sniper rifles (both .308 and .338 Lapua) and FN Minimi machine guns (both 5.56 and 7.62). The rifles were equipped with Schmidt & Bender 3–12x50 scopes and the MGs with Aimpoint Micro T-1 and 3XMag. Everyone was issued enough ammo to shoot each weapon available. Targets from were spread from 100 to 800 meters. The sniper rifles available included HK417 (20-inch bbl), the amazing Timberwolf from PGW Defence Technologies, and Sig Sauer's SSG 3000. When the line went cold, a former Swedish SOF guy from Taiga Climate Protection Systems introduced us to their line of clothing products to keep the body warm and dry in any weather conditions.

In the evening, there was another live fire demonstration. Nammo showed us the power of their 5.56 M955 and 7.62 M993 AP ammo by firing at armor steel plates (12mm and 18mm respectively) that were pierced like butter. At night, we watched their IR tracers in action. The Brügger & Thomet representative performed a demonstration of their less lethal system, composed of their LL-06 40mm launcher and less lethal ammo (OC, CS, foam, rubber, marker, smoke) and gave us the chance to shoot an impact round. The representative from Rheinmetall shot a selection of 40mm grenades (day and night markers, door breachers, IR tracer, smoke, HE grenades, and stun grenades). As night fell, it was our time to shoot with night vision to check out the new Aimpoint LPI (a really small and effective IR laser). We then went through a night recon exercise arranged by SureFire.

Thursday, September 2

Time to shoot bigger weapons! We started the morning firing some training rounds with the Carl Gustav 84mm recoilless rifle produced by Saab Bofors Dynamics. I even had the opportunity to fire an 84mm round; it was amazing, no recoil at all. Aiming and hitting the target were really easy with the Aimpoint FCS12.

Then I took the FN .50 caliber M2 machine gun equipped with the Aimpoint MPS3 heavy weapon sight. I dialed in the range on the sight's ballistic compensator and landed a few bursts of lead on target. Easy and effective!

We then moved on to 40mm grenade launchers. First a few rounds from the HK GMG were fired aiming with the Aimpoint FCS12, and we easily hit targets 500 meters away. Next I used an HK AG36 (the FN40GL-S was also available) to fire some Rheinmetall grenades. I felt the thump on my chest from the increased recoil of the medium velocity rounds (800-meter effective range). Hitting the target was easy with the Micro T-1 on the quadrant mount from Spuhr. This mount even had tritium inserts so that I could visually check range settings at night.

In the afternoon, we fired some handguns, personal defense weapons (PDWs), and ARs against Canaxa pop-up targets and DART-X Range Robots.   The Peltor guys provided ComTac II and Maxim eye protection.

Friday, September 3

Unfortunately, this was the last day of the event. This range day was really worthwhile. I had the good fortune of spending an exciting few days of trigger time with a great group of people in a gorgeous setting and learned about a number of new and exciting products that make our missions easier to accomplish. As an added bonus, before leaving Ravlunda, we got to pick the Aimpoint sight of our choice. I chose the CompM4s. For a Spanish Marine, that was an unexpected surprise at the end of an event to remember!

Also Available from Skyhorse Publishing

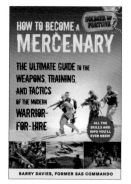

Along with *Soldier of Fortune Magazine*, Barry Davies teaches you the training and knowledge that goes into being a mercenary. Davies will also go into the history of the profession and show how it has evolved. It's always been about the money, but in this book, you will learn all the skills that you must acquire before you take your first job.

$16.99 Paperback • ISBN 978-1-5107-5542-0

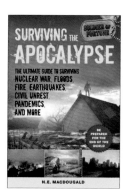

Being prepared for what's out there is important—you have to know what to do when everything falls apart. Knowing how to survive will prepare you for anything and everything that could possibly go wrong. From packing the proper survival kit, to surviving on the battlefield, being physically fit, and coping in the event of a socio-economic collapse, *Soldier of Fortune Magazine*, along with N. E. MacDougald, will make sure that you're never caught off-guard.

$16.99 Paperback • ISBN 978-1-5107-5268-9

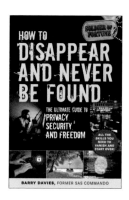

Whether you're being followed or need to get away as soon as possible, being able to disappear without a trace is something that you will need to know. Whether you're an ordinary civilian or a military operative, having this skill is imperative to ultimate survival . . . and there's nobody better at knowing how to vanish at a moment's notice than a former SAS expert. Barry Davies has solutions to all of these problems and more with the *Soldier of Fortune Guide to How to Disappear and Never Be Found*.

$16.99 Paperback • ISBN 978-1-5107-5267-2